Managing Diversity and Ii
Real Estate Sector

Research shows that high-performing organisations focus on diversity and inclusion (D&I). In any workplace, it is important to both understand and recognise the benefits that having a D&I workforce provides. It is integral to developing people within an organisation, serving clients as best we can, and playing an important leadership role in communities.

This book is the first to place D&I at the centre of successful real estate and construction organisations. It provides guidance to, and most importantly, actions for professionals in the sector who want to make D&I an inherent part of the culture of their organisation. This book has been written to bring the sector up to speed with what D&I is all about and how a D&I strategy can be implemented to secure future success. It presents a practical and easy-to-read guide that can help organisations and their leaders engage with and apply this agenda to win the war for talent in real estate and construction.

This book is essential reading for all property leaders and professionals working in the real estate and construction sectors. Readers will gain especially from personal reflections on all aspects of diversity by a broad range of people working in the property industry.

Amanda Clack was the 135th President of the Royal Institution of Chartered Surveyors (RICS). She is an Executive Director at CBRE where she is Head of Strategic Advisory and Managing Director for Advisory across EMEA. She sits on the CBRE UK Board. She has an MSc in Programme Management and is a Fellow of RICS, the Institution of Civil Engineers, the Association of Project Management, the Institute of Consulting and the Royal Society of Arts. Amanda is a Certified Management Consultant and Companion of the Institute of Management.

Judith Gabler has worked for RICS in Europe for over 20 years and is currently Acting Managing Director, Europe with Germany as her country focus. She has a BA (Hons) in French with German and an MSc in Real Estate and Property Management. She is also a Fellow of the Chartered Institute of Linguists and the Chartered Management Institute, as well as being a Chartered Linguist and a Chartered Manager.

Managing Diversity and Inclusion in the Real Estate Sector

Amanda Clack MSc BSc PPRICS FRICS FICE FAPM FRSA CCMI

FIC CMC Affiliate ICAEW

Judith Gabler MSc BA (Hons) Dip RSA FCMI CMgr FCIL CL

CBRE ⚫RICS

Routledge
Taylor & Francis Group
LONDON AND NEW YORK

First published 2019
by Routledge
2 Park Square, Milton Park, Abingdon, Oxon OX14 4RN

and by Routledge
52 Vanderbilt Avenue, New York, NY 10017

Routledge is an imprint of the Taylor & Francis Group, an informa business

This book is adapted from an earlier German language version of the work:
Diversity & Inclusion Management in der Immobilienbranche: Ein
Praxisguide by Amanda Clack, Judith Gabler and Maarten Vermeulen
Copyright © Springer Gabler 2017. All Rights Reserved.

British Library Cataloguing in Publication Data
A catalogue record for this book is available from the British Library

Library of Congress Cataloging-in-Publication Data
Names: Clack, Amanda, author. | Gabler, Judith, author.
Title: Managing diversity and inclusion in the real estate sector / Amanda
    Clack and Judith Gabler.
Description: Abingdon, Oxon ; New York, NY : Routledge, 2019.
Identifiers: LCCN 2018047484| ISBN 9781138368903 (hbk) | ISBN
    9781138368910 (pbk) | ISBN 9780429428975 (ebk)
Subjects: LCSH: Real estate business—Management. | Diversity in the
    workplace.
Classification: LCC HD1375 .C585 2019 | DDC 333.33068/3—dc23
LC record available at https://lccn.loc.gov/2018047484

ISBN: 978-1-138-36890-3 (hbk)
ISBN: 978-1-138-36891-0 (pbk)
ISBN: 978-0-429-42897-5 (ebk)

Typeset in Times New Roman
by Swales & Willis Ltd, Exeter, Devon, UK

Illustrations by Farley Katz, *The New Yorker*. All illustrations are copyright
CBRE Ltd

MIX
Paper from
responsible sources
FSC
www.fsc.org
FSC™ C013985

Printed in the United Kingdom
by Henry Ling Limited

Thank you for purchasing our book. All the authors' proceeds from the sale of this book will be passed on to the Registered Charity LandAid as a contribution to help end youth homelessness in the UK as part of RICS' Pledge to help raise funds in its 150th anniversary year.

Amanda Clack and Judith Gabler

LandAid
THE PROPERTY INDUSTRY CHARITY

uniting to end youth homelessness

Every year, thousands of young people become homeless. Many are forced to sleep in overcrowded hostels, on people's sofas or even on the streets. LandAid is the property industry charity working to end youth homelessness in the UK. It brings together remarkable businesses and individuals across the property industry to support projects providing life-changing accommodation for young people facing homelessness. By awarding grants for building works and arranging free property expertise, LandAid enables charities supporting young people to renovate existing or build new, suitable, safe and affordable accommodation, helping thousands of young people to reach their potential.

We dedicate this book to all those trying to make a difference to diversity and inclusion in the workplace, no matter where they sit in the organisational hierarchy, and to those who have supported us in trying to "do our bit", but in particular:
Uwe Schoessow
and
Peter and Olive Clack

"Here's to DiversiTEA and INKlusion!"

Contents

Acknowledgements

A sincere thank you to both **Dr Sean Tompkins** and **Martin J. Brühl** for their generosity in lending their names and support by writing the forewords to this publication.

We would particularly like to thank two exemplary organisations in the real estate and construction sector, **CBRE** and **RICS**, for their input and unwavering support, without which this publication would not have been possible.

We are particularly indebted to all the amazing people who have shared their very personal and humble life stories. Their perspectives come from their origins around the world, with countries represented such as the Caribbean, India, Iran, Kenya, South Africa, Spain and the UK. We acknowledge with humility the individual journey they have all travelled and the strength it has given them to share the pain and the joy of that journey with others. Their messages for the CEO are therefore in particular both poignant and uplifting.

Kinga Barchon MRICS, Partner, PwC Poland

Arun Batra LLB (Hons), CEO/Founder UK National Equality Standard

Antonia Belcher BSc (Hons) Dip Arb FRICS FCI Arb MAPM C. Build E FCABE FRSA, Founding Partner mhbc

Ciaran Bird, UK Managing Director, CBRE Ltd

Justin Carty BSc (Hons) MRICS, Senior Director, CBRE Capital Advisors Ltd

Kimberly Hepburn AssocRICS, Junior Quantity Surveyor, Transport for London

Amy Leader BSc (Hons) MRICS, Associate Project Manager, Oxbury Chartered Surveyors

Pinky Lilani CBE DL, Founder and Chair of Women of the Future Awards and the Asia Women of Achievement Awards

Antonio Llano Batarrita API MRICS, Llano Realtors SL

Hashi Mohamed MSc BSc, Barrister at No5 Chambers and broadcaster at the BBC

Ali Parsa MPhil, PhD, Professor of Real Estate and Land Management, Royal Agricultural University

Gian Power BSc (Hons) ACA, Founder and CEO of TLC Lions and Unwind London

Sanett Uys MBA MRICS MD, Serendipityremix

Ashley Wheaton BA (Hons), Principal, University College of Estate Management (UCEM)

Sue Willcock MBA (Construction and Real Estate) Dip Proj Man (RICS) Assoc CIPD, Director, Chaseville Consulting Ltd

Our first book, *Diversity & Inclusion Management in der Immobilienbranche: Ein Praxisguide*, was co-authored together with **Maarten Vermeulen** FRICS and published in October 2017 for the German-speaking markets. We would like to warmly thank him for his unwavering support in allowing us to continue without his contribution as co-author for this English edition.

Foreword: Diversity in action

Our sector has to adapt to the combined seismic forces of markets, geopolitics, digitalisation and urbanisation. This at a time when demand for natural resources is outstripping supply, our climate is heating beyond sustainable levels, and we face a demographic double whammy of a burgeoning but ageing global population.

We can rise to these challenges by harnessing the best of technology and the best of human ingenuity: disruption not evolution. As public interest bodies, professional institutions must lead the way in thought and deed. The professions have a responsibility to take the long-term view, seeking evidence from multiple perspectives, and developing solutions that are creative and readily scalable.

The approach outlined above is a definition of diversity in action.

I am convinced that an inclusive mind-set is essential to how we train and equip professionals in times of rapid change. But let us face some hard realities: women still bear a disproportionate share of care for children and elderly family members. Traditional attitudes and dogma do not change overnight, so if we want women to fulfil their potential in our sector, as leaders, we must lead the way from the boardroom in terms of practical steps to help and set up our organisations to embrace diversity and inclusion because this has to start at the top.

At RICS, we are committed to the removal of barriers, establishing flexible working practices, new career paths, and overcoming the gender pay gap. We are ensuring under-represented groups are seen to have equal status in our profession, and we are tackling unconscious bias. Today – with the technology and resources available to us – there should be no excuse for a person's gender, sexuality, race or creed preventing them from making their full contribution. I congratulate the authors in continuing to champion this vital issue.

Dr Sean Tompkins DPhil (Hon) DEng (Hon) ACII FCIM
FIDM CCMI FIOD
Chief Executive Officer
RICS

Foreword: Inclusion is not an illusion

The real estate sector is in the throes of change and facing manifold challenges. An "overheated" market cycle protracted by money politics, more stringent regulatory regimes as a result of the unfettered financial crisis, as well as multiple new sources of global uncertainly push the management and staff of real estate organisations to the limits each and every day, and in every part of the value chain.

In times when retrospective-looking populism is penetrating the geographical boundaries and mindsets of nation states, and when the value of human work being done by digital artificial intelligence is being questioned, then we have to put the human capital we have been entrusted with to the best possible use in order to overcome the present challenges: We are all stewards responsible for our planet.

You will see from reading this book that we can still do quite a lot in our notoriously conservative sector to literally get the "best" out of human interaction by:

- overcoming ingrained images of society;
- releasing creativity by opening up new perspectives;
- encouraging talent and resourcefulness in a collaborative organisation.

Promoting diversity and inclusion in the real estate sector was a topic close to my heart as the 134th RICS President in 2015–2016. I am delighted to see how successfully the authors have since taken this topic to the next level.

With this in mind, I am wholeheartedly recommending this book to you. Let's collectively show how erroneous one slogan of the most recent US presidential campaign is and prove that for our sector, inclusion is *not* an illusion.

Martin J. Brühl FRICS
Chief Investment Officer
Union Investment Real Estate GmbH

Introduction

Research clearly shows that high-performing organisations are diverse and inclusive. In any workplace, it is important to both understand and recognise the benefits that having a diverse and inclusive workforce provides. Diversity and inclusion (D&I) is integral to developing people within the organisation, to serving clients in the best possible way, as well as to playing a leadership role in communities. Essentially it is about valuing everyone in the organisation as an individual and enabling them to be themselves and perform at their best.

Embracing D&I recognises the value that difference brings in terms of considering opinions, perspectives and cultural references. Promoting and supporting D&I in the workplace is not just about good people management, it is much more. To truly realise the benefits from implementing D&I in the organisation, it is vital to create an inclusive environment. As authors, it is our belief that only an organisation that respects and encourages D&I will ultimately succeed in the global marketplace.

Diversity is about differences and individuality. We must recognise that each of us is different and that it is important to value and respect individual differences such as gender, ethnicity, nationality, age, background, education, working and thinking styles, as well as religious background, sexual orientation, ability and technical skills.

Inclusion is about creating an environment where differences are embraced and where all people feel, and are, valued – where they can bring their differences to work each day, and where they can contribute their personal best in every encounter.

A D&I workplace is essentially an environment where everyone feels a part of the whole no matter who they are. It allows people to be the best that they can be and offers a place where everyone feels able to participate and realise their potential. D&I must therefore become an inherent part of the cultural DNA of the organisation. Behaviours that do not conform need to be constructively challenged and amended, alongside D&I being openly discussed to encourage better outcomes.

A diverse team brings the best perspectives and ensures the best possible alignment with client and societal norms. This is an essential ingredient in any sector that is largely a service-based people environment, and not just the real estate and construction sector. For example, boards that have a balance of female and male directors show a commitment to diversity that goes beyond just checking the box.

D&I needs to be both visible and felt within the organisation. It has to start at the top, with the CEO, but then permeate and filter throughout the whole organisation.

This book sets out some of the leading practice and thinking in this area, which can be complex in a global context. It has been written as a guide, ideally for the CEO or those in leadership roles within an organisation who are in a position to move the D&I organisational dial. It explains why and how to implement a vision and policy for D&I in the context of a real estate and construction environment. But it has also been written for anybody who wants to become informed concerning what D&I is all about – the overall principles can be applied to many organisations across the globe, irrespective of size or sector, taking into account local laws as well as cultural, religious and societal factors.

The book is intended to be neither exhaustive nor unduly brief, but aims to provide a perspective taking some of the theory and practice into meaningful action. Our single ambition is to help the organisation, but especially the CEO, leaders and contributors to the agenda, to help move the dial, if only even a little, for real estate and construction.

We hope that our practical and easy-to-read guide will help you start straight away to move the dial on your agenda and make a difference in your organisation, whatever role you may be in. We hope you feel inspired and convinced by the fact that D&I must be an integrated part of your overall organisational strategy in order to remain relevant in a world that becomes more global every day. Put simply, addressing D&I and integrating it into your business strategy will positively affect your bottom line, make your employees value your business differently and will reflect the values of your organisation to your clients. But most importantly in today's world, it is, quite simply, the right thing to do.

Everybody, not just the CEO, can play a leading role in changing the culture in the organisation. D&I is the responsibility of everyone. Embracing D&I not only makes business sense, it also makes for a positive corporate culture. Ultimately it ensures that we can attract top talent not just to the real estate and construction sector, but to any sector anywhere in the world.

1 Setting the scene

1.1 Introduction

For many organisations D&I is synonymous with being part of a corporate social responsibility (CSR) agenda. Across the globe a growing number of industry leaders are reinforcing their commitment to embracing a culture, both internally and externally, of principles and values of which D&I is just one part of the whole. The United Nations Global Compact[1] stands firm on commitments from over 9,867 companies in 164 countries to embed into their overall strategies, culture and day-to-day operations ten universal principles of sustainable business practices covering human rights, labour, the environment and anti-corruption. Worldwide, real estate and construction organisations are also aligning their business practices with the principles of CSR in order to recruit, develop and retain an increasingly diverse and inclusive workforce. By doing so they are establishing core business practices that are considered essential ethical choices, positively defining the organisational reputation and establishing the bedrock for long-term sustainable business models.

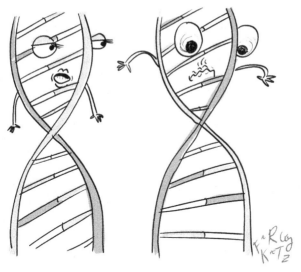

"Your culture is showing."

1.2 A global challenge

If many organisations are already interacting within their own sphere of influence, it seems out of place that many others (still) feel challenged by D&I. One of the main reasons for this is that the road to understanding and embedding D&I

*"This Rubix cube is multi-faceted, multi-layered and multi-complex.
Let's just smash it and glue it back together."*

is multi-faceted, multi-layered and multi-complex. Secondly, D&I is continually evolving, as organisations themselves evolve and develop, with established cultural norms being challenged in support of the war for talent, equality and the creation of a balanced and harmonious workplace. Centuries of pre-conditioning, culture, socio-economic evolution and demographics, to name but a few, mean that a true D&I culture will, for many, not land overnight with a bang, but will be a gentle evolution of step-by-step progress. In reality it can take time to move the organisational dial, but it does gain greater momentum when the collective power of all stakeholders working together helps make that change more quickly.

1.3 Legislation

Underpinning CSR are multiple layers of legislation, guidelines and codices. While these create a strong foundation upon which to embed a D&I culture, they are in turn bound by restrictions.

Take India, for example. As a union of states it is governed by the overarching Constitution of India,[2] which first came into effect in 1950, but which is regularly updated in order to include any new legislation. Accordingly, all citizens have a right to equality before law with the State prohibiting discrimination on grounds of religion, race, caste, sex or place of birth and stipulating equal pay for equal work for both men and women.[3] The Rights of Persons with Disabilities Act was also passed in India in 2016[4] so that it could meet its obligations to the United Nations Convention on the Rights of Persons with Disabilities,[5] alongside the United Kingdom, the USA and many European countries, and which it had already ratified in 2007. Yet the sheer size of a rapidly growing 1.35 billion population together with huge social, economic and environment problems mean that change is extremely slow. In March 2017, for example, the *India National Herald*[6] raised its voice against 15 different forms of direct and indirect discrimination, including biased recruitment practices due to caste, harassment relating to gender identity, and segregation due to religion.

Yet there is hope. While metamorphosing out of long-standing traditions and customs takes a seemingly endless amount of time, awareness-raising and lobbying, change can nevertheless happen. The most recent example is the landmark ruling by the Supreme Court of India in September 2018 which overturned colonial-era legislation known as section 377 and which now allows sex in private between consenting homosexual adults.[7] It is a beacon of encouragement that will hopefully set an example for others to follow.

In Great Britain, more than 110 individual pieces of legislation are combined into the Equality Act 2010[8] in order to protect people in the workplace and in wider society. It is a comprehensive piece of legislation applicable throughout England, Scotland and Wales setting out who is protected from discrimination, the different types of discrimination, and action that individuals can take if they feel they have been discriminated against. The nine "protected characteristics" are discussed in more detail in Chapter 2.

In Continental Europe it is the European Union (EU) that is responsible for forging political and economic collaboration, maintaining stability and peace, and creating a European identity in a globalised world. Currently with 28 member countries and 24 official languages, it is founded on the universal values of human dignity, freedom, equality and solidarity so that all are "united in diversity". Human rights are embedded in the Charter of Fundamental Rights of the European Union[9] and are legally binding in the EU since the Treaty of Lisbon was enacted in 2009. European law enforces a number of directives in order to protect citizens' rights against discrimination, and all the D&I characteristics are covered within the individual categories of laws both in and beyond the workplace. Directive 2000/78/EC[10] in particular establishes and upholds the overall framework for non-discriminatory practices in the workplace. At national level, member states and non-EU member states have in turn interpreted and implemented the directives within local legislation.

Yet while legislation can provide a framework to protect individuals' rights to be treated fairly, equally and, where required, anonymously, it can also mean that data collection and monitoring of progress can be (considerably) restricted.

In order to better assess the impact of data legislation, the Royal Institution of Chartered Surveyors (RICS) commissioned research in late 2017 into four countries in Continental Europe. This was important given the cultural diversity and potential sensitivities around D&I in some countries about what is "acceptable" or "not acceptable" to openly discuss and disclose. The research was a legal assessment[11] of existing legislation in France, Germany, The Netherlands and Spain relating to six characteristics of age, gender, ethnicity, disability, religion and belief, as well as sexual orientation. Parental leave and data around socio-economic background were also considered as a secondary focus. The primary objective was to identify those questions specifically relating to the six characteristics that a third party could legally ask an organisation to disclose based on six elements of the workforce life-cycle, namely leadership, recruitment, development, retention, engagement and continuous improvement. The three main conclusions indicated that:

1   In all four countries there are no restrictions on asking firms to provide general information relating to the organisation's contact person, sector and number of employees.

2   There are also no restrictions on asking an organisation whether:

   a   there is a D&I strategy in place;
   b   D&I is a corporate goal underpinned by objective-driven action plans;
   c   the organisation has a data protection policy;
   d   D&I related data is collected.

3   There are restrictions in three of the four countries when it comes to workforce monitoring. Data collection and processing can only be effected with prior standard consent in Germany, France and Spain due to strict limitations

of the data protection legislation and especially due to the perceived risk that data provided anonymously may become inadvertently linked to individual employees. In these countries:

a   An organisation can disclose whether it collects data relating to age, gender, ethnicity, disability, religion and belief, sexual orientation and parental leave.

b   An organisation may not disclose specific data provided by the workforce self-identifying either gender (e.g. male, female, transgender), ethnicity, physical or mental impairment, religion, sexual orientation, or parental leave inasmuch as the latter relates to gender.

In the mean time, the introduction of the EU General Data Protection Regulation (GDPR)[12] in May 2018 has fundamentally changed how data can be collected, stored and administrated. While the legislation goes a considerable step further in protecting the rights of the data subject, including data access and the right to be forgotten, for the purpose of D&I monitoring and benchmarking it has certainly added a further layer of complexity.

Government legislation is a key anchor to changing the D&I landscape globally, but the transformational journey is not without roadblocks. Years of culture and tradition may impact legislation. Consequently, data collection, while providing indicators of workforce composition and how this changes over a period of time, may be restricted in scope and depth by what is legally permissible.

"Watch out, INKlusion—that old outdated law is going to put up a fight."

1.4 International firms versus SMEs

The rich fabric of diversity poses different challenges for different organisations simply because of varying size, purpose or sector. An international organisation with business operations and subsidiary offices in several global markets has different resources than, for example, a micro, small or medium-sized enterprise (SME) with headcounts of < 10, < 50 and < 250 respectively. Larger organisations will certainly have easier access to best-practice sharing and will most likely have a different sense of urgency given the size of the workforce.

Many SMEs struggle with D&I. They grasp the overall concept and the added value it can give, but are challenged by how it can be implemented on a day-to-day basis in an all-consuming resource-constrained, profit-driven and highly competitive environment. Micro or small companies can certainly consider a step-by-step evolution approach. For example, every management team, however small, can act as a role model by encouraging and displaying inclusive behaviours. By optimising the use of digital technology, it is also possible in today's modern environment to create the facilities for an organisation of any size to be able to offer remote working options if or when required. This is particularly important in providing workplace flexibility in order to support childcare or employees who may have other caring responsibilities (such as elderly or ailing parents or siblings). These small steps usually do not involve any major resource or cost investment, but have far greater motivational impact when the employee can work without worrying about what is happening "back home" or "back in the office".

SMEs in Europe also have the opportunity to join the network of the EU Platform of Diversity Charters.[13] The platform was set up in 2010 and is funded by the EU Commission in order to encourage and support organisations of all sizes at an individual member state level. There are currently 21 European Diversity Charters in place. By signing a "charter", organisations publicly commit to a set of principles and are united in one common understanding, to share best practice and implement diversity-related activities such as workshops and networking events. Since the platform transcends organisational size, geography and sector, it is a good source of inspiration for ideas and learning beyond the real estate and construction sector.

"On close inspection—that is an authentic mustache."

In general, however, all companies, irrespective of size, need to pay attention to how they publicly position their commitment to D&I. Authenticity is an imperative – if an organisation publicly promotes D&I on its website homepage, then employees expect that the D&I culture will be genuinely embraced and lived by that organisation. They will be quick to be critical of hollow or shallow publicity or marketing that do not correspond to the realities of the workplace. Disengagement, and ultimately resignation, can otherwise be the ultimate consequences.

1.5 Our inherent cultural DNA

A diverse and inclusive culture can also be achieved if the principal stakeholders are willing to critically assess their current business practices, to adapt these, and to visibly advocate change within their sphere of influence. It is inevitable that multicultural differences prevail at world regional, sub-regional and country level. They are evident through social backgrounds, the way we think, the way we speak, the way we express ourselves, our heritage, upbringing, our values and our economic status. The combinations are endless. The following section takes a brief look at culture, language and behaviour as intrinsic elements of our individual DNA, and why we therefore need to recognise these in ourselves and in others when trying to create diverse and inclusive workplace environments.

"I'm holding two stakes, so I'm basically a principal stakeholder."

Culture

Culture formed by the environments we grow up in, as well as years of tradition and doing things in a certain way, pre-condition the way we think. It therefore goes without saying that D&I initiatives cannot be replicated with the same level of success in every region of the world or every organisation. While Western societies may focus on aspects such as disability, sexual orientation and gender, in other world regions, such as Asia, D&I may be related more to caste as well as regionalism, language and language dialects, and these in turn are linked to social mobility.

At the workplace, people of the same nationality or religion may naturally gravitate to one another because they share the same values or heritage, although for others this may be perceived as forming a "closed group". On the other hand, people who have worked across borders and cultures tend to appreciate differences better and adapt this understanding when working internally across teams, or externally with clients, tenants and communities. So hiring people who have worked in different jurisdictions and geographies is an effective way of creating a culturally versatile talent pool for the organisation and helping to break down stereotypes and prejudices.

"Just because I come from a place where hats are large and you come from the land of small hats doesn't mean we can't work together."

Language

In the UK the ability to communicate is relatively simple given that English is the predominant native language. Nevertheless, regional dialects and accents are prevalent, and over the years a melting pot of linguistic diversity has been greatly enhanced by guest workers and migration. London is now host to the largest number of community languages in Europe, with over 300 languages being spoken. The 50-plus countries of Continental Europe share between them 24 official languages as determined by the European Union, with Switzerland alone laying claim to four of these – German, French, Italian and Romansh – as national languages. In India the official government languages are Hindi as well as English, but there are in total 22 officially recognised languages as determined by the individual states.

So language skills and multilingualism are essential features of cultural diversity and competence, and have been proven to have additional psychological and lifestyle advantages as well as to boost the economy and earning power. The Chartered Institute of Linguists,[14] for example, is committed to international understanding and respect for the diversity of languages and cultures, and while there is far more to be said about these as well as the social, economic and cognitive benefits of languages, for the purpose of D&I it is clear that the culturally enhanced multilingual individual is well placed for success.

Nevertheless, organisations struggle to recognise these benefits, and despite growing internationalisation, setting an official language policy within an organisation can be both challenging and costly, with many organisations opting for English coupled with selective translation of priority documents and policies depending on the topic and audience. While this may be customary practice and makes business sense from a cost and resource perspective, it can mean that words and meanings are literally "lost in translation", resulting in minority language groups within an organisation feeling isolated or excluded from a perceived privileged inner circle.

Successful organisations, as businesses, therefore need to recognise, value and develop the linguistic skills of their employees. They must ensure that they can speak the language of the market they will be operating in, providing support and training, as well as ensuring that there is an adequate budget to facilitate this.

Adoption of an inclusive written medium has yet to be established as a norm in business and within organisations. Perhaps not surprisingly, the extremely conservative Académie française in its role as guardian of the French language has opposed the use of gender-neutral job titles and has even issued a declaration condemning the use of gender-neutral writing as it would put the French language *en péril mortel* (in mortal danger).[15] Inevitably, this is a discussion that will most likely be continued in years ahead as customs and practices of culturally acceptable norms evolve and change.

Behaviour

Another aspect to consider is human behaviour. Research by McKinsey[16] has established that lessons can be learnt from behavioural economics and social psychology whereby "human behaviour is heavily influenced by subconscious, instinctive, and emotional 'System 1' responses, rather than being under the exclusive control of rational, deliberate 'System 2' thinking". Negative terminology often associated with D&I behaviours and mindset barriers includes words such as close-mindedness, discrimination, harassment, victimisation, favouritism, homogeneity bias, labelling, negativity and prejudice, as well as subconscious or unconscious bias. This can result in stereotyping, for example with regard to gender or sexuality, whereby individuals are pigeonholed into the expectation of only being able, or competent, to do certain jobs or types of work. Attitudes and behaviours are therefore heavily influenced by cognitive biases that affect decision-making.

"It pains me to admit this, but my cognitive bias has led me to prejudge you as a hot head."

This is further evident since women have traditionally been stereotyped as non-assertive, so while a male leader may be seen as directive, the stereotyped equivalent of a women doing the same job could most likely be classed as bossy, arrogant and authoritarian.

1.6 Message to the CEO

With this brief introduction to CSR and the elements of D&I, we hope to have shown you the foundation for sustainable business practices as well as engendering diverse and inclusive teams at the workplace. The CEO is in a unique position to be the catalyst for questions that people are otherwise afraid to ask in their organisations by visibly opening up conversations and taking an interest in diverse subjects such as how people celebrate different religious festivals and what these mean to them. This leads to the spirit of organisational sharing, learning and inclusiveness. The CEO can create the culture for a learning, insightful and enquiring organisation, and give permission for it to be part of the DNA.

In this spirit, you may like to consider the following:

1   **Consider your own attitude and that of your senior management team** to CSR and how, within that context, D&I business practices can be embedded into the organisational culture.

2   **Inform yourself of current legislation and any "protected characteristics" in the countries in which your organisation (and clients) operate.** Consider how you can integrate these into your D&I strategy (see Chapter 6). It is important to be aware of data legislation that can support, but may also restrict, data collection and monitoring.

3   **If you are CEO of a micro or small company with limited resources, consider how to leverage what you have** – such as making use of digital technology to help create a more flexible workplace, or identifying opportunities

to collaborate with others in order to become part of a wider or global D&I community and learn from leading practice examples. Size should not stop any organisation from creating an inclusive workplace.

4   **Acknowledge culture, language and behaviours as the chief stimuli for creating diversity and improving inclusivity in the organisation.** Be open-minded about national and regional cultures and the "why". Deliberately hire people who are organisationally culturally different, and create culturally diverse talent pools that in turn will foster open-mindedness and tolerance. Encourage workshops and roundtables – verbal communication can often achieve more than a written policy and break down stereotyped thinking and prejudices.

5   **Support your employees in gaining cultural knowledge and experience** – for example, empower your human resources (HR) teams to set up personal development systems whereby talent may rotate at different geographical locations. While each person will make their own cultural assessment based on subjective experience, the most valuable assets are to be geographically flexible and to be open-minded enough to understand and appreciate that people are different, both in their private and working life. One way of developing cultural (and linguistic) competence is "country hopping", whether this be a personal career choice or whether the CEO is willing to fund staff development via a secondment, for example.

6   **Review your language policies** and, where possible, encourage multi-lingualism and language learning. Embed these in recruitment and people development policies that enhance your business with a multilingual task force.

7   **Encourage and reward behaviours and attitudes linked to D&I,** such as tolerance, empathy, open-mindedness, agility and flexibility, curiosity and looking beyond the label. Make your people feel part of the culture shift by asking them to share best-practice examples from their own country that can be adopted within the inclusive workplace.

8   **Conduct a 360-degree reflection of all the stakeholders** and identify any useful collaboration partners, focusing on gender, supporting new talent entering the profession, mixed generation teams, etc.

9   **Finally, remember to be authentic – if you say it, do it.**

Notes

1 United Nations (2018). United Nations Global Compact. Available from: www.unglo balcompact.org/ [Accessed 30 September 2018].
2 The Constitution of India (1950). Available from: www.india.gov.in/my-government/ constitution-india/ [Accessed 30 September 2018].
3 The Constitution of India (1950), §§14–16; §39(d). Available from: www.india.gov.in/ my-government/constitution-india/ [Accessed 30 September 2018].
4 National Centre for Promotion of Employment of Disabled People (2016). Rights of Persons with Disabilities Act. Available from: www.ncpedp.org/RPWDact2016 [Accessed 30 September 2018].

5 United Nations (2006). United Nations Convention on the Rights of Persons with Disabilities. Available from: www.un.org/development/desa/disabilities/convention-on-the-rights-of-persons-with-disabilities.html [Accessed 30 September 2018].

6 NH Political Bureau (17 March 2017). 15 examples of discrimination in India. *National Herald*. Available from: www.nationalheraldindia.com/news/a-new-bill-against-discrimination-lists-common-forms-check-out-if-you-know-of-citizens-subjected-to-discrimination [Accessed 30 September 2018].

7 Safi, M. (6 September 2018). Campaigners celebrate as India decriminalises homosexuality. *The Guardian*. Available from: www.theguardian.com/world/2018/sep/06/indian-supreme-court-decriminalises-homosexuality [Accessed 30 September 2018].

8 Equality Act 2010. Available from: www.legislation.gov.uk/ukpga/2010/15/contents [Accessed 30 September 2018].

9 Eur-Lex (2016). Charter of Fundamental Rights of the European Union, 2016/C 202/02. Available from: https://eur-lex.europa.eu/legal-content/EN/TXT/?uri=celex:12016P/TXT [Accessed 30 September 2018].

10 Eur-Lex (27 November 2000). Council Directive 2000/78/EC. Available from: https://eur-lex.europa.eu/LexUriServ/LexUriServ.do?uri=CELEX:32000L0078:EN:HTML [Accessed 30 September 2018].

11 Conducted by RICS and DLA Piper.

12 EUGDRP.ORG (2018). EU General Data Protection Regulation – GDPR. Available from: https://eugdpr.org/ [Accessed 30 September 2018].

13 European Commission. EU Platform of Diversity Charters. Available from: https://ec.europa.eu/info/policies/justice-and-fundamental-rights/combatting-discrimination/tackling-discrimination/diversity-management/eu-platform-diversity-charters_en [Accessed 30 September 2018].

14 www.ciol.org.uk [Accessed 30 September 2018].

15 Académie française (2017). *Déclaration de l'Académie française sur l'écriture dite "inclusive"*. Available from: www.academie-francaise.fr/actualites/declaration-de-lacademie-francaise-sur-lecriture-dite-inclusive [Accessed 30 September 2018].

16 McKinsey & Company (2015). *Diversity Matters*. Available from: www.mckinsey.com/~/media/mckinsey/business%20functions/organization/our%20insights/why%20diversity%20matters/diversity%20matters.ashx [Accessed 30 September 2018].

2 The relevance of D&I in real estate and construction

2.1 Introduction

D&I has become increasingly prominent on boardroom agendas as the war for talent has grown ever more acute. The need to attract, develop and retain talent is key to business survival, and organisations around the world are having to open the gate wider, while maintaining a bar, to attract people from a wider and more diverse pool. Nowhere is this more acute than in the real estate and construction professions, where the land, construction and property sector has been slow to react and adapt to the working demographic.

Looking to the growing future demand for talent to serve the growing needs of the built environment in delivering the buildings, infrastructure and places for businesses and people to exist in, we need to also consider the role of the professions and other key stakeholder groups alongside the employer organisations, and understand the role they can play in attracting, educating, qualifying and developing talent.

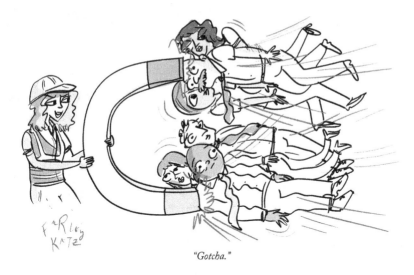

"Gotcha."

2.2 Our changing world: let's be ready

RICS undertook a global survey in 2014, *RICS Futures*, which took place in over 19 countries across the world. The final report, *Our Changing World: Let's Be Ready*, has been very influential in terms of developments and trends the built environment will face in the foreseeable future.[1] From this research, six areas for focus were identified because their impact will be most significant:

1   winning the war for talent;
2   ethics at the heart of everything we do;
3   smart and sustainable cities;
4   technology and big data;
5   the increasing importance of infrastructure;
6   leadership.

Each of these topics offers great opportunities for growth and development, but also challenges which need to be overcome – challenges that largely tie in with making sure that, now and in the future, the land, construction and property sector will be able to attract enough talent for the work that needs to be done. Key areas considered here are the importance of leadership and winning the war for talent in tackling the importance of D&I within organisations.

What is most interesting from this research is that the primary concern of the employers interviewed was how to attract and retain talent in their workforce, both now and in the future, while the number one concern of the professionals interviewed was how they can remain relevant in a rapidly changing world. The war for talent emerged as a key finding to address these concerns, and within the battle for talent, the need to recognise the whole of society – and hence the increasing importance and awareness of embracing a diverse workforce and considering

"This means war! War for talent. Sorry, I didn't mean to alarm you."

greater inclusivity to optimise the widest possible opportunities for employment. This matters in a global context because of the scale of the demand, and therefore the opportunity pool, for professionals in the land, construction and property sector to help address the increasing societal demands for skills. The following trends and figures may help to bring this to life:[2]

- By 2050 there will be 6.3 billion people living in cities – so urban expansion will reach significant heights and densities, requiring more homes and city development to meet the rapidly expanding attraction of the city as a place to work, live and spend recreation time.
- There will be a US$57 trillion infrastructure gap to fill by 2030 – with countries like the US needing to invest US$1 trillion in the next ten years. Plus 50% of the world's infrastructure required by 2070 will need to be built, requiring huge global construction programmes to meet the demand.
- By 2030 the world will need 50% more food, 45% more power and 30% more water than in 2012. As a result, there will be increasing pressure on the world's natural resources, meaning that land and resources will need more effective management.

These figures clearly show that the demand for talented people will increase significantly, and the land, construction and property sector needs to develop talent by embracing greater D&I in the workforce in order to meet these requirements. Failure to do so will simply mean that organisations will be looking in a smaller pool of talent for the workforce they need now and in the future.

"I want a bigger talent pool. And some swim trunks."

This is particularly heightened at entry level, because the sector is often not seen as very attractive by school children, those entering further and higher education, graduates or other potential pools of labour. This is why initiatives to take the message of opportunities in this sector need to be spread further down into schools to attract the professionals of the future.

"There's no such thing as starting too early."

The answer to this challenge is to start developing a sector that is seen as diverse and inclusive, and therefore will be capable of attracting more people from a wider proportion of society.

Moreover, the Chartered Management Institute in London has estimated that by 2025, improving D&I could add as much as US$12 trillion annually (US$75 billion in 2017) to the global economy.[3] So it is clear that addressing aspects of D&I makes economic and business sense as we engage in the war for talent to help service the increasing demands for skilled professionals for real estate and construction in particular.

The scope of the profession itself is incredibly diverse, including land, minerals, construction, infrastructure, the environment, real estate, facilities management, asset management and fine art, to name but a few. The fact that it is so diverse means there are many opportunities, given the globally increasing demand for such professionals, but especially for women, who at 50% of the world's population are still significantly under-represented at all professional levels, but particularly in this sector at senior levels in organisations.

As of 2018, 14% of RICS qualified members were female and 24% were graduates and trainees – however, in areas such as quantity surveying or construction more broadly, this figure was well below 6%.[4]

The land, construction and property sector is still conspicuously male-dominated, and cannot continue with so few senior women in executive leadership positions. Bringing in more women to the sector will open up opportunities for greater profits and returns for investors, as leading research by EY and the Property Council of Australia shows.[5] In other words, companies need to move from "We know gender diversity is important" to "We understand and embody in our attitudes and behaviours the business imperative of gender diversity for our industry."

"Practise what you preach, my brother!"

It goes without saying that D&I extends beyond the gender debate: it needs to be considered for all aspects of diversity, based on the understanding and inclusion of those with "protected characteristics". Protected characteristics vary from country to country, but generally these refer to groups such as the nine pro-tected under the Equality Act 2010,[6] set out by the Equality and Human Rights Commission in England and Wales:

1   age (and generational differences);
2   disability;
3   gender reassignment;
4   marriage and civil partnership;
5   pregnancy and maternity;
6   race or ethnicity – BAME (Black, Asian or Minority Ethnic);
7   religion or belief;
8   sex;
9   sexual orientation.

To improve the sector's approach to D&I, there is a need to improve understanding and why embracing D&I in the workplace both opens up the potential for a better working environment, but also makes business sense in terms of attracting the best talent to the organisation, developing the best people and driving the bottom line.

Of course, a level of interpretation is important when looking at the protected characteristics, as these can also be heterogeneous. What is important is not to think of these rigidly, but in a nuanced way when considering D&I and how it should apply to the organisation.

External pressures are there, too, including societal ones, but essentially people want to work in an environment that allows for, and actively embraces, D&I.

Diverse people = diverse thoughts

This is essential when it comes to attracting top graduate talent as well as retaining more senior talent.

Talent is in demand. To ensure companies get the best selection of the best talent, they need to be fishing in the widest possible pool of people, otherwise they will be trapped in a vacuum of self-reliance and static monolithic thinking. Diversity of thought comes with a diversity of people. The power to embrace D&I within the cultural DNA comes from effective (thought) leadership.

2.3 The role of RICS in D&I

RICS is the global professional body that actively promotes trust in the profession. With its core focus on the public good, RICS works to make the land, construction and property sector more ethical, professional and transparent through four clearly defined and interlinked activities:

1   **International co-development of standards** – Due to the globalising world and significantly increasing cross-border investments, global players seek consistency in many ways. RICS provides that consistency through technical industry guidance and by co-developing international standards related to ethics, valuation, property, construction and land measurement.
2   **Accreditation** – To make sure that RICS professionals are able to turn theory into practice, they need to be assessed and show that they are qualified to do the job. Accreditation is the standard all RICS professionals have to achieve to gain and maintain membership.
3   **Education and training** – Once standards have been developed, people need to be trained in how to apply them. RICS provides relevant education and training within the context of a broader training offer in order to facilitate life-long learning.
4   **Regulation** – Every RICS professional needs to comply with the RICS Code of Conduct, and failure to do so may result in being sanctioned, which can vary from having to pay a fine to losing accreditation. Regulation offers assurance to the market and the consumer, and is for the public good.

As a professional body, RICS strongly feels the responsibility to be a thought leader and trend watcher, and an initiator when it comes to taking new directions and leading the industry. The *RICS Futures* research speaks for itself, but also, from an ethical perspective, it is justified and morally right to treat everybody equally. The land, construction and property sector is relatively conservative and somewhat resistant to change. Therefore, a professional body like RICS, with its primary purpose to act in the public interest, is ideally positioned to start and fuel the discussion around D&I. In many ways, RICS has become an industry leader when it comes to engendering D&I at all levels across the profession, and as an employer itself. This ranges from encouraging recruitment of the next generation from as wide a catchment as possible (inclusion) to working with organisations in a programme to look at aspects of diversity in tackling the war for talent.

2.4 Other key influencers and stakeholders on the D&I agenda

Fortunately, despite the many challenges, numerous individuals, organisations and governments are working together with the common goal of making the D&I change happen. This section will briefly examine some of the main stakeholder groups and what they are already doing to create inclusive and diverse workplaces:

1   international organisations;
2   real-estate industry bodies and branch organisations;
3   D&I organisations;
4   education providers.

International organisations

For the purpose of this subsection, international organisations are defined as those entities that have a global reach and influence, transcending any particular sector. They may represent one particular group or may be multi-faceted, but collectively these international organisations will be committed in general to creating more diverse and inclusive cultures.

Leading the way as an exemplary organisation is the United Nations (and its associated agencies, such as the International Labour Organization), serving to create better workplaces, or the United Nations Educational, Scientific and Cultural Organization (UNESCO), which promotes, among other things, the equal dignity of all cultures, education and gender equality. UN International Days centrepiece gender equality (Women's Day), disabilities (Day of Persons with Disabilities) and racial diversity (Day for the Elimination of Racial Discrimination) and celebrate, for example, multilingualism (International Mother Language Day) and heralding for 2019 a World Day for Cultural Diversity. Such international days sharpen global awareness and acceptance of differences in society and at the workplace.[7]

Real estate industry bodies and branch organisations

Historically the land, construction and property sector has not been particularly renowned for its innovation, but it is now becoming increasingly receptive to, and is trying out, more and more methods of talent management that involve D&I principles. Industry bodies and branch organisations are in a unique position to exert influence. Depending on their constitutional status, real estate bodies and organisations are usually outward-looking in order to serve specific markets or act in the public interest, rather than being driven by profit or commercial interests. They are therefore in a good position to catalyse discussions around D&I and encourage cross-sharing within and across industries, either via their member organisations, their members, their collaboration network with other organisations, or individuals in leading or influential positions.

Alongside RICS, many other organisations are extending their portfolios in order to improve D&I globally. Two examples from the USA are the Urban Land Institute's Women's Leadership Initiative programme[8] and the Building Owners and Managers Association International, which was one of the first to set up a diversity committee in 1994 for the Chicago region in order to provide scholarships and make education more accessible to minorities.[9]

D&I (networking) organisations

There are, in addition, many organisations that are solely committed to raising D&I within the land, construction and property sector, driving change and removing barriers through research, benchmarking studies, communication and dialogue. Some of these focus on improving gender equality and have been in existence for a number of years. They commonly offer a range of networking and professional development opportunities, provide access to research and case studies, showcase diversity champions and role models, and achieve high-profile visibility via events, conferences and awards. These include Catalyst in the USA; Commercial Real Estate Women (CREW), also in the USA; Women in Property (WIP) in the UK and associated networks across Europe; the Women's Property Network in South Africa; and Women of the Future in South Asia and the UK. There are, of course, many more, and a more comprehensive overview will be covered in Chapter 3. However, the fact that there are so many indicates a coming together of like-minded individuals supporting one another to help make a difference and to provide encouragement, role models and information to others in the sector.

Education providers

Change is accelerating, and our global world is becoming increasingly complex, hence roles and skills within the built environment are also changing. Having an international education or international awareness is extremely important, and there is an increasing demand for international courses in different languages with opportunities to combine study terms in different countries. Quality courses help shape and mould the professionals of tomorrow, with new specialisms and different formats – such as online learning, flexible learning for full-timers, part-timers, senior experts or young students – whatever their preference or prior learning may be.

Education providers play a pivotal role – besides influencing and inspiring students and alumni, they can appeal to employers to provide interesting job opportunities, and they can prompt students to futureproof their skills. They can serve as academic role models by designing attractive teaching curricula that embrace strong leadership, offer insight into intercultural competence, international standards, professional ethics, and embrace the use of multilingual teaching and teaching locations. This will in turn drive the number of graduates entering the real estate profession upwards.

Employers can work with education providers across geographical borders to create a culturally resilient environment. The International Real Estate Challenge (IREC),[10] for example, is a global initiative that brings together real estate students from different universities and countries in a common learning and development mission. Launched in 1999, it has engaged over 700 students from Europe, Ireland and more recently Washington, DC. Each year, pre-selected students are tasked with creating a fictitious solution for a real-life client that is then assessed after two weeks of site tours and project groups by an expert jury of industry practitioners.

Such initiatives are not just about real estate challenges, though, they are also about providing opportunities to push the boundaries of intellectual, cultural and learning capabilities and even being part of a global real estate community. In its 150th year, RICS launched the Cities for our Future[11] challenge in partnership with UNESCO and the Association of Commonwealth Universities, inviting students from the global built environment to compete for a £50,000 global cash prize by proposing innovative ideas and solutions to the most pressing global challenges, such as rapid urbanisation, resource scarcity or climate change.

Higher education alumni can also play a key role in building bridges between education and the real estate market. Greater levels of D&I can be achieved by demonstrating to students the wide range of attractive careers available and by helping them to develop more confidence in their own skills. Role models play a critical leadership role, and many organisations are working hard to get more women, especially at a senior level, into the public eye. University alumni organisations can facilitate integrative working by providing a platform for students to meet with experienced graduates via networking events, mentoring programmes or workshops. There are many examples of real estate alumni networks which operate at global and/or regional level, such as the Technische Universität Wien in Austria with its womenTUsuccess initiative,[12] which showcased 25 female professionals as examples of the impressive diversity of attractive real estate career paths.

2.5 Message to the CEO

What is clear is that RICS takes very seriously its responsibility to actively engage with its stakeholders on D&I on a global basis. Part of this role involves offering guidance to the industry based on identified developments and trends while always acting in the public interest. This has also led to developing a repository of leading practice, developing a data set based on the RICS Inclusive Employer Quality Mark (IEQM) benchmark (see Chapter 5) and dedicating a page on the RICS website to D&I.[13] The website specifically details the following as part of building a 21st-century professional body through equality, driven by D&I:

> The adoption, promotion and embedding of a culture of trust and diversity is a priority for RICS and diversity must become a core strand of all RICS thinking and business strategy.

Respecting and valuing differences, with a commitment to equal treatment and equal opportunity, will help us serve the public, our professionals and regulated firms effectively, and deliver policies and outcomes which are fair and transparent as well as to ensure that our profession is fit and relevant for the future.

Diversity and inclusion is a significant subject area, not just for RICS, but for society as a whole. There are many issues to be tackled and many stereotypes to be broken down. In order to be successful, we must focus on those areas where we can be effective and deliver the greatest degree of change.

The work of the RICS as a regulatory body has an impact on millions of people . . . . As a result, the diversity of people we come into contact with, and offer services to, is immense. We therefore need to fully incorporate equality into the way we do business, making every effort to eliminate discrimination, promote equal opportunities and help our staff and professionals to be the best they can be.

As a CEO, having an understanding of the various types of "protected characteristics" is essential. The ability to adapt and flex these according to the global nature of your business and the legal jurisdictions of each country require specialist HR and legal advice. It is important to understand that some characteristics still cannot even be discussed legally in many countries. Being cognisant of the wider aspects of "protected characteristics" is an essential starting point, coupled with an understanding of what can and cannot be covered within specific countries in which your organisation operates. However, it is imperative to set an agenda that is inclusive and embraces the opportunities that a diverse workforce can bring for an organisation – but most importantly, for clients and staff.

"Adapt! Flex! Are you Chief Executive Officer? Come on, get in shape!"

It is also important to have a wider awareness of other key stakeholders, or influencers, who can help contribute and lend support to the organisational D&I journey. For your organisation, developing a strategy that positively includes D&I is necessary to meet various stakeholder demands, but most importantly, to make sure that your organisation has a sustainable future when it comes to having the right people to serve (future) market and client demands.

Notes

1 RICS (2015). *Our Changing World: Let's Be Ready*. Available from: www.rics.org/ globalassets/rics-website/media/knowledge/research/insights/rics-futures-our-chang ing-world.pdf. [Accessed 30 September 2018].
2 RICS (2015). *Our Changing World: Let's Be Ready*. Available from: www.rics.org/ globalassets/rics-website/media/knowledge/research/insights/rics-futures-our-chang ing-world.pdf. [Accessed 30 September 2018].
3 Chartered Management Institute (CMI). Gender parity could boost global GDP by $28 trillion. Available from: www.managers.org.uk/insights/news/2016/february/gender-parity-could-boost-global-gdp-by-twentyeight-trillion [Accessed 30 September 2018].
4 RICS. How can we attract and retain more female talent? Available from: www. rics.org/uk/wbef/markets-geopolitics/how-can-we-attract-and-retain-more-female-talent/ [Accessed 30 September 2018].
5 Property Council of Australia (2016). Property Council and EY release ground break-ing gender profile report. Available from: www.propertycouncil.com.au/Web/Content/ Media_Release/National/2016/Property_Council_and_EY_release_ground_break ing_gender_profile_report.aspx [Accessed 30 September 2018].
6 Equality Act 2010. Available from: www.legislation.gov.uk/ukpga/2010/15/contents [Accessed 30 September 2018].
7 United Nations. United Nations observances. Available from: www.un.org/en/sections/ observances/united-nations-observances/ [Accessed 30 September 2018].
8 Urban Land Institute (ULI) (2018). Women's Leadership Initiative. Available from: https://americas.uli.org/programs/leadership-network/womens-initiative/ [Accessed 30 September 2018].
9 Sudo, C. (23 February 2017). How one man helped found the only BOMA diversity program in the country. *Forbes*. Available from: www.forbes.com/sites/ bisnow/2017/02/23/how-one-man-helped-found-the-only-boma-diversity-program-in-the-country/#74592a4a24d7 [Accessed 30 September 2018].
10 International Real Estate Challenge (2018). Available from: http://irec-global.net/start. html [Accessed 30 September 2018].
11 RICS (2018). Cities for Our Future Challenge. Available from: www.citiesforourfuture. com/ [Accessed 30 September 2018].
12 Technische Universität Wien (2013). womenTUsuccess. Available from: www.tuwien. ac.at/dle/genderkompetenz/best_practice_projekte/womentusuccess/ [Accessed 30 September 2018].
13 RICS (2018). Diversity and inclusion. Available from: www.rics.org/eu/about-rics/ responsible-business/diversity-and-inclusion/ [Accessed 30 September 2018].

3 Gender parity

3.1 Introduction

A truly diverse and inclusive working environment respects all human differences. While this book has been written to review all aspects of D&I in its widest possible sense, in terms of the organisation tackling the war for talent for the benefit of bringing the best people into the organisation, there is no question that with statistics for female qualified RICS professionals being at just 14% of a global membership in excess of 100,000, we would be remiss to omit a chapter on gender specifically. Improving gender balance is key to tackling the war for talent, considering that the World Bank has found that 49.5% of the total population and close to 39.5% of the working population are female. The sector, professions and organisation need to do more to increase the number of women, especially, in the world of land, construction and real estate.

3.2 The gender gap

In 2006 the World Economic Forum launched the annual *Global Gender Gap Report* in order to raise global awareness of the challenges posed by the gender gap, and stimulate debate, solutions and actions. The report benchmarks countries based on the four sub-indices of economic participation and opportunity, educational attainment, health and survival, and political empowerment. *The Global Gender Gap Report 2017*[1] analysed and ranked 144 countries against these four factors and concluded that while many countries had made considerable progress towards achieving gender parity, globally it was shifting into reverse that year for the first time since 2006, with the average progress on closing the global gender gap standing at 68.0% – an average gap of 32.0%, compared to 31.7% in 2016. The gender gap, as measured by the World Economic Forum, is the difference between women and men as reflected in social, political, intellectual, cultural, or economic attainments or attitudes.[2]

The roots of the gender gap, and therefore gender inequality, go far and deep. It is extremely difficult to achieve the same pace of change in every country or world region simply because there are many civilisations where certain behaviours and beliefs have been ingrained over centuries. Behaviours that may be accepted as the cultural or social norm are inevitably handed down from generation to generation. It is always important to remember that many people do not have a choice of career due to their social background, wider responsibilities, and family and other social pressures. Aspects of gender inequality have been created and have evolved for many social, economic and political reasons, in order to ensure lineage and inheritance and to maintain power.

In the UK, women had to wait until the Representation of the People Act in 1918 to be allowed to vote – but only if they were over 30 and met certain conditions of property ownership. However, it was not until the Equal Franchise Act was enacted in 1929 that the same rights were given to both men and women over the age of 21. In countries such as Afghanistan, female access to education is

Table 3.1 Overview of gender gap per world region

Region	Gender gap	Time needed to close the overall gender gap
Western Europe	25%	61 years
North America	28%	168 years
Eastern Europe and Central Asia	29%	128 years
Latin America and the Caribbean	29.8%	79 years
East Asia and the Pacific Region	31.7%	161 years
Sub-Saharan Africa	32.4%	102 years
South Asia	34%	62 years
Middle East and North Africa	40%	157 years

Source: Adapted from *The Global Gender Gap Report 2017.*[3]

way below the norm, in North Sudan or Zambia land ownership is still restricted, and in many religions child marriages and/or honour killings are still practised in order to preserve lineage. In Saudi Arabia, from June 2018, women are no longer banned from driving, yet there are many other examples from across the globe where a level of gender restriction still exists whereby some of these practices are enshrined in laws and are likely to take years to change.

However, through peer, political and social pressure putting the spotlight on these factors, change will gradually come. Gender parity still has a long way to go, but it is one of the highest priorities simply because it is not just about equal rights for everyone, but bringing a whole new workforce of capable women into the workplace.

In a working environment, probably the most common examples of gender inequality relate to access to jobs (recruitment, job adverts, internal), remuneration, workplace flexibility and promotion opportunities. There are numerous initiatives around the world to help change this, and in the following section we explore in more detail the following areas:

1   the female quota;
2   gender pay reporting;
3   closing the gender pay gap.

The female quota?

Another challenge to the real estate and construction professions is to increase the number of women as role models, on boards and in leadership positions.

Whether it is effective or not to introduce a quota per se has already been debated by many reliable sources. Research by Credit Suisse[4] has made the case for more female diversity at senior and management level across all sectors. Based on a total analysis of 27,000 senior managers at 3,380 companies globally, only 61 companies demonstrated that women account for 50% and above of senior management ("the 50% club").

As we have seen in Chapter 2, the European Commission is working hard to redress the balance in gender equality, and legislation has been introduced in a number of countries over the past 10–15 years. In November 2017 *The Guardian*[5] reported that the EU is pushing for a 40% quota for women on company boards, which exceeds the 30% benchmark which came into effect in Germany on 1 May 2015[6] for 30% female representation in any new supervisory board positions of listed companies.

Many find the concept of quotas unpalatable, and many female leaders who have made it to the board room would be uncomfortable with the concept that they got there simply as a result of the organisation achieving its quota target, rather than being there as a result of merit.

The alternative is, of course, setting targets rather than quotas. A target sets out an aspirational direction for the organisation based on key metrics such as graduate intake, recruitment, promotions or gender (or other) representation in the board room, but still allows the organisation to grow organically based on a wider set of merit-based criteria while also challenging itself to properly consider and evaluate decision-making based on the aspiration to achieve a broader-based diversity representation.

"I'm all for setting targets, but are we insured for this?"

A key question remains about how women can get more support and increase their visibility so that they can penetrate a predominantly male environment. In Germany, for instance, the Institute for Corporate Governance in the German Real Estate Industry (ICG)[7] has introduced a "Diversity/Women on Boards" mentoring programme for equal- or mixed-gender teams or tandems in order to prepare high-potential women for board or supervisory board functions in three to five years.

What is key for the organisation is to properly challenge the rationale, stereotypes, groupthink and unconscious bias against embracing a more gender-diverse (and broader) workforce.

Gender pay reporting

The gender pay gap is now reported on in many different countries of the world, and is the difference in the average hourly wage of all men and women across a workforce. If women do more of the less well-paid jobs within an organisation than men, the gender pay gap is usually bigger. The gender pay gap is not the same as unequal pay, which is paying men and women differently for performing the same (or similar) work. In countries, such as the UK, for example, unequal pay has been unlawful since 1970.

According to the World Economic Forum, two-thirds of the Organisation for Economic Co-operation and Development (OECD) countries have introduced policies on pay equality since 2013, including requiring some employers to publish calculations every year showing the gender pay gap.[8] Gender pay gap reporting is now delivered in many countries across the world, but it is particularly notable in Europe, where countries, such as the UK in 2018, have made gender pay gap reporting mandatory: employers with 250 or more employees must publish figures comparing men and women's average pay across the organisation. However, on 1 January 2018, Iceland became the first country in the world to pass legislation with real teeth outlawing the gender pay gap, including a requirement on companies and government agencies that employ more than 25 people to obtain a government certificate demonstrating pay parity, or else they will face fines.[9]

Despite these initiatives, however, the global gender pay gap is widening. Globally, women earn about 57% of what men earn, which means that the gap between men and women will not be closed for another 217 years.[10]

"If we're both holding stakes to keep watch for vampires,
I better be getting paid the same."

Offering flexibility can be a double-edged sword when it comes to gender pay gap reporting. For example, part-time work is typically done by a higher proportion of women than men, which can impact an organisation's structural hierarchy when it comes to reporting numbers.

If not for altruism's sake, leaders around the world have a financial incentive to take a note of what Iceland has done and work towards closing the gender pay gap in their countries. Economic gender parity could add an additional $250 billion to the GDP of the United Kingdom, $1,750 billion to that of the United States, $550 billion to Japan's, $320 billion to France's, and $310 billion to the GDP of Germany. Furthermore: "the world as a whole could increase global GDP by $5.3 trillion by 2025 if it closed the gender gap in economic participation by 25% over the same period".[11]

In many countries, such as the UK, large employers are legally required to publish gender pay gap data on their own websites and on the government website.

The rules are slightly different for employees in the public sector and those in the private and voluntary sectors. Employers with 250 or more employees must calculate and publish their data as follows:

- mean gender pay gap;
- median gender pay gap;
- mean bonus gender pay gap;
- median bonus gender pay gap;
- the proportion of men in the organisation receiving a bonus payment;
- the proportion of women in the organisation receiving a bonus payment;
- the proportion of men and women in each quartile pay band.

Closing the gender pay gap

The UK Government gender pay gap reporting[12] points to some useful actions to consider for organisations looking at closing the gender pay gap. These fall into two categories: (a) effective actions and (b) promising actions.

Effective actions

Actions such as these have been tested in real world settings and found to have a positive impact:

- **Include multiple women in shortlists for recruitment and promotions** – when putting together a shortlist of qualified candidates, make sure that more than one woman is included. Shortlists with only one woman do not increase the chance of a woman being selected.
- **Use skill-based assessment tasks in recruitment** – rather than relying only on interviews, ask candidates to perform tasks they would be expected to carry out in the role they are applying for. Use their performance on those tasks to assess their suitability for the role. Standardise the tasks and how they are scored to ensure fairness across candidates.
- **Use structured interviews for recruitment and promotions** – both structured and unstructured interviews have strengths and weaknesses, but unstructured interviews are more likely to allow unfair bias to creep in and influence decisions. Use structured interviews that ask exactly the same questions of all candidates in a predetermined order and format, and grade the responses using pre-specified, standardised criteria. This makes the responses comparable and reduces the impact of unconscious bias.
- **Encourage salary negotiation by showing salary ranges** – recognise that women are less likely to negotiate their pay. This is partly because women are put off if they are not sure about what a reasonable offer is. Employers should clearly communicate the salary range on offer for a role to encourage women to negotiate their salary. This helps applicants know what they can

reasonably expect. If the salary for a role is negotiable, employers should state this clearly as this can also encourage women to negotiate. If women negotiate their salaries more frequently, they will end up with salaries that more closely match those of men.

- **Introduce transparency to promotion, pay and reward processes** – transparency means being open about processes, policies and criteria for decision-making. This means employees are clear what is involved, and that managers understand that their decisions need to be objective and evidence-based because others can review them. Introducing transparency to promotion, pay and reward processes can reduce pay inequalities.
- **Appoint diversity managers and/or diversity task forces** – diversity managers and task forces monitor talent management processes (such as recruitment or promotions) and diversity within the organisation. They can reduce biased decisions in recruitment and promotion because people who make decisions know that their decisions may be reviewed. This account-ability can improve the representation of women in the organisation. Diversity managers should:
  - have a senior/executive role within the organisation;
  - have visibility of internal data;
  - be in the position to ask for more information on why decisions were made;
  - be empowered to develop and implement diversity strategies and policies.

Promising actions

These actions are promising, but require further research to improve the evidence about their effectiveness and how best to implement them:

- **Improve workplace flexibility for men and women** – advertise and offer all jobs as having flexible working options, such as part-time work, remote working, job-sharing or compressed hours.
- **Allow people to work flexibly, where possible** – encourage senior leaders to role-model and champion flexible working. Encourage men to work flex-ibly as well, so that it is not seen as only a female benefit. Avoid flexism, whereby the organisation, and managers in particular, exhibit characteristics that discriminate against those who desire to work flexibly, such as resent-ment, mistrust and ultimately "ganging up" against them because they are seen as being treated specially. Encourage flexibility around performance being about outputs, not presenteeism.
- **Encourage the uptake of shared parental leave** – the gender pay gap widens dramatically after women have children, but this could be reduced if men and women were able to share childcare more equally. For example, in the UK shared parental leave and pay enable working parents to share up to 50 weeks

of leave and up to 37 weeks of pay in their child's first year. Offer enhanced shared parental pay at the same level as enhanced maternity pay, and encourage the take-up of shared parental leave.

- **Recruit returners** – returners are people who have taken an extended career break for caring or other reasons and who are either not currently employed or are working in roles for which they are over-qualified. Key considerations for recruiting returners include targeting places where returners are likely to be looking for work, ensuring the recruitment process is returner-friendly, and offering support before and during the assessment.
- **Offer mentoring and sponsorship** – although they are quite similar roles, mentors provide guidance and advice to their mentees, while sponsors support the advancement and visibility of those they are sponsoring. Some evidence suggests that mentoring programmes work very well for some women, but not for others. It is not clear, based on existing evidence, whether sponsorships are more effective than mentoring, or how best to run mentoring and sponsorship programmes so they are effective.
- **Offer networking programmes** – some evidence suggests that formal networking programmes where people meet and share information and career advice can be helpful for some women, but not others. More work is needed to understand the effects of networking programmes and whether they need to have particular features in order to be successful.
- **Set internal targets** – it is important to ensure that employers' equality goals are clear and realistic, and that progress towards them can be tracked. "Improving gender equality at my organisation" or "reducing my organisation's gender pay gap" can be overarching goals, but they are not specific, so they risk being unsuccessful. One way of increasing the likelihood that goals will be reached is by setting specific, time-bound targets: what change will be achieved, and by when.

3.3 Gender equality

As explained in the subsections on gender pay reporting and closing the gender pay gap above, equal pay is not the same as gender pay gap reporting. Yet equality of pay between men and women in many countries still remains an issue. A recent study by the World Bank utilising data from 141 countries found that gender equality would enrich the global economy by an estimated US$160 trillion if women were earning as much as men in the workplace.[13] Effectively, equal pay, equal hours and equal participation in the workforce could lead to a global wealth jump of US$23,620 per person, as well as creating knock-on benefits such as lower malnutrition and child mortality rates.

Researchers found that countries are losing 14% of their wealth, on average, simply because of gender inequality. The World Bank economist and report author, Quentin Wodon, said:

By looking at 141 countries, which is most of the world, we could see that basically everywhere women are earning less than men. So we calculated how much more wealth there would be, worldwide, if women were earning the same as men for the same wage, and the same hours worked.

It therefore seems that regardless of intent, regulation or legislation, equal pay overall still remains an issue.

The *2017 Report on Equality between Men and Women in the EU* by the European Commission[14] shows that Europe has come a long way in ensuring greater equality between women and men. Vigilance remains key on this issue, it seems, as in the context of persistent economic inequality and rising intolerance – both online and in the public sphere – it is essential for the EU to reaffirm its strong commitment to gender equality. The principle of equality between women and men has been enshrined in the EU treaties since 1957. Since then, much progress has been made, with women gaining the benefits of better education and thereby having been able to increase their presence in the labour market. The importance of equality between women and men is a fundamental value of the EU, and one that has been enshrined in treaties from the very beginning, as the Treaty of Rome included a provision on equal pay. Over the last 60 years, societal changes and persistent policy efforts have established a growing trend towards gender equality.

For Europe, at least, the EU has always been a major force behind these developments, with its annual report contributing to the monitoring and in-depth review of the Sustainable Development Goal (SDG) on gender equality of the UN 2030 Agenda and of some other SDGs, including indicators with a gender perspective.[15] SDG 5 on gender equality is captured in the box below.

Sustainable Development Goal 5 targets

a) End all forms of discrimination against all women and girls everywhere.

b) Eliminate all forms of violence against all women and girls in the public and private spheres, including trafficking and sexual and other types of exploitation.

c) Eliminate all harmful practices, such as child, early and forced marriage and female genital mutilation.

d) Recognise and value unpaid care and domestic work through the provision of public services, infrastructure and social protection policies and the promotion of shared responsibility within the household and the family as nationally appropriate.

e) Ensure women's full and effective participation and equal opportunities for leadership at all levels of decision-making in political, economic and public life.

f) Ensure universal access to sexual and reproductive health and repro-
    ductive rights as agreed in accordance with the Programme of Action
    of the International Conference on Population and Development and
    the Beijing Platform for Action and the outcome documents of their
    review conferences.

g) Undertake reforms to give women equal rights to economic
    resources, as well as access to ownership and control over land and
    other forms of property, financial services, inheritance and natural
    resources, in accordance with national laws.

h) Enhance the use of enabling technology, in particular information
    and communications technology, to promote the empowerment of
    women.

i) Adopt and strengthen sound policies and enforceable legislation
    for the promotion of gender equality and the empowerment of all
    women and girls at all levels.

3.4 Changing gender perceptions by the next generation

In 2016, RICS undertook a survey with YouGov, an international Internet-based
market research and data analytics firm headquartered in the UK, of females and
males aged of 13–22. The results from the survey say a lot about what we need to
do to help move the dial on perceptions around gender.[16]

The survey identified that the most diverse industries were perceived as being
retail and health, with law and construction seen as the least diverse. What was
striking was that of the number of females interviewed, almost 30% considered
construction as being only for men.

The survey also identified that a quarter of young women believed they
would do better under the leadership of a female CEO and that they wanted
to see more visible role models. The rise of political leaders around the world
not only gave many inspiration about the potential they could achieve, but they
also felt it would help change sexist attitudes and help encourage workplace
diversity – with 43% believing that having a female prime minister or president
would encourage gender diversity at work. But it was surprising that in 2016,
41% of these young women still believed their gender would hold them back in
the workplace.

Young men seem to think differently, with 20% saying that they expect to earn
more in their careers than their female counterparts.

Finally, a striking 72% of those surveyed believed that the attitudes and behaviours of CEOs and senior leaders are important in encouraging equal numbers of
men and women.

Theolodite parity achieved!

What is clear is that gender matters, and we have a long way to go to myth-bust engrained stereotypes, encourage more visible role models and help influence this next generation that real estate, and construction, is a place for them, offering an exciting career, equal opportunities and an improving pay gap at all levels.

"The future is bright. And not just because of my lottery ticket."

3.5 Creating the right environment

A report by the Urban Land Institute[17] reviewed women in real estate through the lens of women in the United States and the Asia Pacific (APAC) region, which are on opposite sides of the globe. These two regions are typically thought to be culturally distinct, and the research indeed highlighted some differences between these two geographies in terms of what women value in the workplace and what they believe they need to achieve professional success. Primary among these differences were the opposing views on the importance of having external relationships. Survey respondents in the United States, and particularly those in the position of CEO, emphasised the significant role that building relationships with industry professionals outside their own organisation played in their success.

More importantly, the report also revealed important similarities shared by women in both regions.

One key similarity pertained to having an employer that is receptive to flexibility-related needs, which goes beyond the need for flexible/generous maternity leave. While respondents to the survey indicated that they need their organisations to formally provide maternity leave, they also expressed that their ability to achieve success is enhanced when their employer formally offers them the opportunity to work on a flexible basis during other periods in their lives.

The key metrics to consider from this research that were identified as being critical to women's professional success were the following:

1   the importance of an inclusive culture;
2   the need for strong relationships, particularly with senior-level leaders who can serve as career sponsors;
3   access to visible and challenging job assignments;
4   the enactment of objective hiring and promotion policies.

"I'm seeing organisational culture and internal relationships."   *"I'm seeing an overpriced crystal ball 0.76m away."*

The organisational culture and internal relationships ranked as two of the most critical informal structures that work to support women's professional success. What was interesting was that access to visible and challenging job assignments, along with the enactment of objective hiring and promotion policies, were slightly more important to US survey respondents, but APAC respondents still highlighted them as key components in supporting their professional success.

Furthermore, women in both geographies expressed ambitious career aspirations, setting their sights on reaching leadership levels in industry organisations. Searching for more responsibility and greater professional growth opportunities could explain why women also show a considerable amount of movement between employers.

This research sends a powerful message to employers about what women need in order to be successful and where employers should focus their resources in order to most effectively support their female employees.

3.6 Female networks

Momentum and awareness of gender parity have also been driven by the growth of female networks – women endorsing the value of other women. Sheryl Sandberg, COO of Facebook, released her first book in 2013, *Lean In: Women, Work, and the Will to Lead*, co-authored with Nell Scovell,[18] which is dedicated "to offering women the on-going inspiration and support to help them achieve their goals". This philosophy has led to the foundation of Lean In and other such networks all over the globe.

The role of female-based networks is to provide an environment for women to come together for professional development, to advance their careers and to provide a forum for them to share common interests. They often provide much-needed support to women, in particular those who operate in a largely male-dominated environment. Sometimes what is key for women in particular is to talk about how they handle various issues in a safe and trusted environment, with Feed In being as important as Lean In for many women.

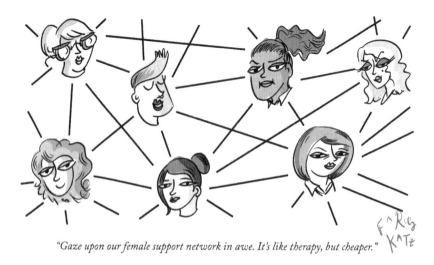

"Gaze upon our female support network in awe. It's like therapy, but cheaper."

Ultimately, we will have achieved our goal in society when there is no longer a perceived need for groups, in this case women, to come together through dedicated networks because these conversations and support happen naturally in the workplace. Until that time, these networks provide an environment for conversation, development and training, often alongside male colleagues.

For real estate and construction, there are equally many networks, too many to list fully. Examples of some powerful female networks for the real estate and construction profession across the globe have been listed below:

- **British Property Federation (BPF)**,[19] founded in 1963, is a not-for-profit membership organisation representing companies involved in property ownership and investment. The BPF "work[s] with Government and regulatory bodies to help the real estate industry grow and thrive, to the benefit of [its] members and the economy as a whole";
- **Catalyst**,[20] founded in 1962 and headquartered in New York;
- **Commercial Real Estate Women (CREW)**,[21] founded in 1989 and headquartered in Kansas as a networking organisation "transforming the real estate industry by advancing women globally";
- **Frauen in der Immobilienwirtschaft e.V.**[22] in Germany;
- **Le Cercle des Femmes de l'Immobilier**[23] in France;
- **Real Estate Balance**,[24] "an association run by a group of female and male leaders from different organisations and disciplines engaged in real estate who are focussed on addressing the gender issue in our sector", because "gender doesn't have an agenda", based in the UK;
- **Salon Real**[25] in Austria for female professionals with diverse backgrounds ranging from architecture, asset management, facility management to agency, valuation, legal and tax advice; like many of its sister networks, it also builds bridges to other European and international networks;
- **Women in Property Switzerland Association (wipswiss)**,[26] which encourages women to have more self-confidence in their own skills, knowledge and professional abilities;
- **Women in Project Management Specific Interest Group (WiPM SIG) of the Association of Project Management**,[27] now in its 25th year, which was founded by a group of women and men to help address the development and promotion of women working in project management environments and works to maximise the availability of expertise in the project management profession;
- **Women in Property (WIP)**,[28] which "creates opportunities, expands knowledge and inspires change for women working in the property and construction industry", based in the UK;
- **Women in Real Estate in Poland (WIREP)**;[29]
- **Women of the Future (WOF)**,[30] a programme for Asia, UK and South Asia founded by Pinky Lilani in 2006, which now has a specific awards category and network for Real Estate, Infrastructure and Construction to "help unlock a culture of kindness and collaboration among leaders" under age 35 plus creating

"collaboration in the workplace, galvanising a community of influential women to work together";

- **Women's Property Network (WPN)**,[31] founded in 2000 in South Africa, "dedicated to advancing the success of women in the Property Industry";
- **World Women in Real Estate (WWIRE)**,[32] which "organises seminars, networking events and technical workshops to give professional women the contacts, knowledge and skill-set necessary for a successful career in real estate".

At national levels, such as in the UK, there are a plethora of networks, including Property Needs You, Changing the Face of Property, Women of the Future, Women Into Construction, Women in Surveying, Real Estate Balance, Women Talk RE, Visible Women and the National Association of Women in Construction – to name but a few!

What is clear is that these types of networks have grown in size, number and influence over the past 10–15 years and have made their voices heard as respected and trusted partners with the ability to inspire, create allegiances and role-model the profession in order to:

- raise the status of women in real estate;
- showcase role models to provide inspiration;
- increase the visibility of women in real estate;
- build potential by connecting people who can share experiences;
- showcase real estate as an attractive career choice;
- demonstrate the ability to combine both a family and career;
- build bridges and strategic alliances by advancing female business partnerships;
- and much more . . . .

3.7 A female perspective: Poland

To obtain a female view of what it is like to be a woman in real estate and construction, we sought one of our first perspectives from Kinga Barchon MRICS, a Partner in PwC Poland. Her contribution helps in considering some of the considerations for women in real estate and construction:

I work in the professional services industry, so our only asset is our people. The war for talent is more and more competitive, with an increasing number of challenges starting from millennials entering and reshaping the workforce, through low representation of women in managerial and leadership positions, ageing workforce and talent pool up to finding the right skills and competencies for expansion into new markets. Therefore, shaping talent strategies to include different talent pools is key for any business to succeed. Diverse talent allows innovation, growth and competitive advantage in the marketplace.

Have women any incentive to become a property professional in Poland?

Starting with the lack of female chief executives, the predominance of men on corporate boards is particularly visible in the property sector. Look as well at the predominance of conference panels consisting of men at all property conferences! But study after study has shown that gender-balanced boards are good for business – improving sales revenue and profit margins. Therefore, the main incentive for a woman in a property business is the possibility to create better outcomes for our clients, our people and our communities. It is proven that diverse team are better placed to serve similar group of customers. We have women in the professional services advising clients, which usually means there is often better communication.

What is the added value that a woman can bring to the property profession?

Currently less than 5% of Fortune 500 CEOs are female. Addressing low representation of women in managerial and leadership positions is key for businesses, as there is little doubt that gender-balanced boards are more successful. Why? The presence of women in a group helps the team to cohere. I perceive this is due to women's social sensitivity awareness, as well as vital listening and bridge-building skills. It is proven that women are better than men in relationship-building, facilitation and empowering others. It results in better productivity and organisational effectiveness. Women drive business models to be more responsive and customer-driven. It is relevant for the property profession, and for many others, to attract more women in!

What sort of practices are firms in Poland adopting to attract a more gender-diverse workforce?

We recognised a long time ago that if we are to solve important problems for our clients, and communities, we need different talent. International firms can develop a globally consistent approach with different initiatives at regional level depending on the market and local needs. But creating the right internal culture is also an imperative that means bringing together individuals of all backgrounds, life experiences, preferences and beliefs. What is most important is recognising that differences can, in fact, be the best business asset you can have.

3.8 Message to the CEO

Later, in Chapter 7, we will look at some thought-provoking personal stories and perceptions that speak to the CEO. As a woman who has had a wonderful career in real estate and construction for over 30 years, one of the co-authors would like to share her own message to the CEO.

Speaking from my own experience, my name is Amanda Clack, and I am the co-author, with Judith Gabler, of this book. People often consider my career as being successful and in many ways high-profile. However, I never set out to put my head above the parapet or to become any form of public role model. What I always wanted was to simply enjoy doing my job, to the best of my ability, to absolutely deliver exceptional service to my clients, and to work with great people around me. I feel very privileged, and incredibly lucky to have had a career that I have loved as much as I do and that has taken me to places I would never have expected – including becoming the 135th President of RICS in the 149th year of the institution, as well as becoming the longest-serving president in 123 years. My full biography is detailed later in the book.

Let me start with some facts: we need to attract more talented individuals into the sector to meet the demand. We need to increase the diversity and inclusiveness of our profession within real estate and construction. This is key, so that we can grow the diversity of talent within our organisations so that when we are putting teams together we are able to select from a wider pool of people and field the best individuals who best represent us to our clients.

I am surprised that, with such great career opportunities that are offered by shaping the world around us in terms of places and property for people, RICS, in its 150th year in 2018, still only had 14% of qualified professionals that were female, albeit at graduate entry level this is now improving somewhat, with female entrants closer to 24%.

But my question is: why are these statistics still so low?

Having joined the profession straight from school, I applied and got into university, but decided instead to take a day release route by studying for my quantity surveying degree part-time over five years. In many ways, it was similar to today's apprentice route. No one in my close family or friends had a professional career in the built environment. Indeed, it was through the encouragement of my parents, my mother in particular, that I considered and selected the surveying profession.

I love this sector and what I do. It is incredible that you can tangibly make a difference to people's lives and the community in the work we deliver through the built environment.

Of course, I am incredibly proud that I became President of the RICS, which was unquestionably a career and personal highlight. To be only the second female president was a huge honour, but I am delighted that I will certainly not be the last. The fact that there are now so many high-profile and successful women in senior and influential roles across the sector proves to me that this is absolutely a career that can be enjoyed by women and in which they can be incredibly successful.

I would say that I have been successful in my career mostly because I love what I do. Has it all been plain sailing? No. Have there been advantages/disadvantages to being a woman in the profession? Yes and no. Do I think this is a career where women can be successful? Absolutely.

My message to you as a CEO would be not to think quota, but to consider the following:

1   Be a role model. Set the vision and standards for D&I that you expect the organisation to adhere to. Go public with organisational targets and a pledge not to accept speaking engagements unless there is a diverse mix, for example.

2   Get the organisational culture right such that it understands and embraces all aspects of D&I at all levels, so that this becomes embedded as part of the organisational DNA.

3   Consider what might work for your organisation to increase gender balance at all levels.

4   What can you do to actively encourage applications from more women? For instance, are your recruitment and promotion policies objective and transparent?

5   How can you personally support women within the organisation, and how can you encourage other leaders to become effective mentors – both when they arrive, and as they go through their professional and personal life?

6   Think about who will be the best candidates for new and challenging assignments and encourage people to think more broadly when putting engagement teams together, giving opportunities to future talent in the organisation and encouraging women to put themselves forward for such roles.

7   Consider setting targets or aspirational objectives for your organisation and the leaders of your business to help move the dial in order to think differently by actively considering the diversity within your organisation. Monitor, measure and publish your data and targets.

Encourage the women in your organisation to be the best that they can be and to aspire to senior positions. This should absolutely not be to the detriment of their male colleagues, but alongside them. For me, it's not either/or, but both!

"Lady teacups—aspire to be big and full of tea like I am!
You too, Coffee Mug Man—come and join the conversation."

There are practical areas where you, and your organisation, can really make a difference by:

- supporting schools initiatives – for example, Class of Your Own in the UK, set up by founder Alison Watson MBE, has been delivering one-day workshops and other built environment programmes to schools and colleges since 2009 to encourage the next generation;
- proactively offering internships and apprenticeships from a wider pool;
- recruiting on the basis of 50:50 male:female, particularly at graduate level;
- supporting women differently in the workplace, as their needs and requirements will invariably be different to their male counterparts';
- helping promote great female role models within the organisation – those who exemplify the right behaviours and values as role models for others to follow;
- ensuring there is diversity on project and assignment teams.

Being a woman in this profession offers an incredibly exciting career, with flexibility and opportunity. All that is needed is for more entrants to come in, succeed and remain in their careers for the gender diversity demographic to change dramatically.

Having more women in senior roles will also help your gender pay gap reporting and will invariably encourage more women to join your organisation.

Gender diversity, as with all aspects of diversity and inclusion, is such as important opportunity for you and the organisation in developing top talent.

It is important to celebrate and consider differences between men and women. Many women need nothing more than being encouraged and supported to speak up and out. Others may need help to push themselves forward in their career. Not being afraid to talk about difference to aid better understanding and to treat all individuals the same, regardless of gender, will provide a better-balanced and effective workforce.

Keep asking what more can you be doing to move the dial to improve the gender balance in your organisation. Not all women are the same, of course, and neither are all men, so understanding and treating your people as individuals, encouraging and celebrating diversity, all makes for a better-balanced and truly diverse working environment.

Notes

1 World Economic Forum (2017). *The Global Gender Gap Report 2017*. Available from: www.weforum.org/reports/the-global-gender-gap-report-2017 [Accessed 30 September 2018].
2 World Economic Forum (2017). What is the gender gap (and why is it getting wider)? Available from: www.weforum.org/agenda/2017/11/the-gender-gap-actually-got-worse-in-2017/ [Accessed 30 September 2018].
3 World Economic Forum (2017). *The Global Gender Gap Report 2017*, pp. vii–viii. Available from: www.weforum.org/reports/the-global-gender-gap-report-2017 [Accessed 30 September 2018].

4 Credit Suisse (2014). The CS Gender 3000: The Reward for Change, pp 27–29. Available from: http://publications.credit-suisse.com/index.cfm/publikationen-shop/research-institute/cs-gender-3000/ [Accessed 30 September 2018].

5 Boffey, D. (20 November 2017). EU to push for 40% quota for women on company boards. *The Guardian*. Available from: www.theguardian.com/world/2017/nov/20/eu-to-push-for-40-quota-for-women-on-company-boards [Accessed 30 September 2018].

6 Bundesministerium der Justiz und für Verbraucherschutz (2015). Meilenstein für Gleichberechtigung: Bundestag beschließt Frauenquote. Available (in German) from: www.bmjv.de/SharedDocs/Artikel/DE/2015/03062015_Frauenquote.html [Accessed 30 September 2018].

7 Institute for Corporate Governance in the German Real Estate Industry (ICG) (2017). AK "Diversity on Boards". Themen. Available from: www.icg-institut.de/themen/ [Accessed 30 September 2018].

8 Harris, B. (2017). What the pay gap between men and women really looks like. *World Economic Forum*. Available from: www.weforum.org/agenda/2017/11/pay-equality-men-women-gender-gap-report-2017/ [Accessed 30 September 2018].

9 Al Jazeera News (2 January 2018). *In Iceland, it's now illegal to pay men more than women*. Available from: www.aljazeera.com/news/2018/01/iceland-country-legalise-equal-pay-180101150054329.html [Accessed 30 September 2018].

10 World Economic Forum (2017). *The Global Gender Gap Report 2017*, p. vii. Available from: www.weforum.org/reports/the-global-gender-gap-report-2017 [Accessed 30 September 2018].

11 World Economic Forum (2017). *The Global Gender Gap Report 2017*, p. viii. Available from: www.weforum.org/reports/the-global-gender-gap-report-2017 [Accessed 30 September 2018].

12 UK Government (2018). Gender pay gap data. Available from: http://gender-pay-gap.service.gov.uk [Accessed 30 September 2018].

13 Hodal, K. (31 May 2018). Gender pay gap costs global economy $160tn, says World Bank study. *The Guardian*. Available from: www.theguardian.com/global-development/2018/may/31/gender-pay-gap-costs-global-economy-160tn-world-bank-study [Accessed 30 September 2018].

14 European Commission (2017). *2017 Report on Equality between Women and Men in the EU*. Available from: https://eeas.europa.eu/sites/eeas/files/2017_report_equality_women_men_in_the_eu_en.pdf [Accessed 30 September 2018].

15 European Commission (2015). The Sustainable Development Goals. Available from https://ec.europa.eu/europeaid/policies/sustainable-development-goals_en [Accessed 30 September 2018].

16 Inman, P. (25 October 2016). Survey: 41% of young women expect to face discrimination at work. *The Guardian*. Available from: www.theguardian.com/money/2016/oct/25/young-women-expect-sex-discrimination-work-royal-institution-charered-surveyors [Accessed 30 September 2018].

17 Urban Land Institute (ULI) (2017). Advancing women in real estate. Available from: https://womeninrealestate.uli.org/ [Accessed 30 September 2018].

18 Sandberg, S. with Scovell, N. (2013). *Lean In: Women, Work, and the Will to Lead*. New York: Alfred A. Knopf.

19 British Property Federation (2018). Available from: www.bpf.org.uk/ [Accessed 30 September 2018].

20 Catalyst (2018). Workplaces that work for women. Available from: www.catalyst.org/ [Accessed 30 September 2018].

21 Commercial Real Estate Women (CREW) (2018). CREW Network: The business advantage. Available from: https://crewnetwork.org/home [Accessed 30 September 2018].

22 Frauen in der Immobilienwirtschaft e.V. (2018). Available from: www.immofrauen.de [Accessed 30 September 2018].

23 Le Cercle des Femmes de l'Immobilier (2018). Available from: www.femmes-immobilier. com/historique [Accessed 30 September 2018].
24 Real Estate Balance (2018). Available from: www.realestatebalance.org/ [Accessed 30 September 2018].
25 Salon Real (2018). Das Netzwerk für Frauen in der Immobilienwirtschaft. Available from: www.salonreal.at/ [Accessed 30 September 2018].
26 wipswiss (2018). Women in Property Switzerland Association. Available from: https:// wipswiss.ch [Accessed 30 September 2018].
27 Women in Project Management (2018). APM Women in Project Management Specific Interest Group (WiPM SIG). Available from: www.apm.org.uk/community/women-in-project-management-sig/ [Accessed 30 September 2018].
28 Women in Property (WIP) (2018). Available from: www.womeninproperty.org.uk/ [Accessed 30 September 2018].
29 WIREP (2018). Women in Real Estate in Poland. http://wirep.pl [Accessed 30 September 2018].
30 Women of the Future (2018). Available from: https://womenofthefuture.co.uk/ [Accessed 30 September 2018].
31 Women's Property Network (2018). Available from: www.wpn.co.za/ [Accessed 30 September 2018].
32 World Women in Real Estate (WWIRE) (2018). Available from: www.wwire.eu/en/ [Accessed 30 September 2018].

4 Workplace realities

4.1 Introduction

The land, construction and property sector is a people business. Organisations thrive and are sustainable when there is a balanced mix of talent, ages and competencies. Multigenerational differences can impact on culture and values, and these can harmonise or polarise, placing different expectations and pressures on the employer.

The challenge is not an easy one: organisations today may have up to four different generations working under one roof. Definitions differ slightly regarding the exact age range of the different generation groups, but these can be generally defined as follows:

- **Baby Boomers** (1946–1964) are more likely to have a high(er) sense of discipline and ambition and have followed a linear career path rather than acquiring a portfolio of projects and experiences. They have acquired their business skills in the first instance via traditional methods rather than having been immersed in technology from adolescence or childhood. Being less

"Trade you some baby food for a bite of that prune stew."

likely to question authority, they may be described as conformist rather than individualistic. This is a generation that lives to work.

- **Gen X** (1965–1976) is more likely to be self-reliant and independent, the work–life balance is important, and the prototype is generally more critical of the stricter "boomer" values of the previous generation, having witnessed first-hand the obligations and restrictions these can impose. This is a generation that works to live.
- For **Millennials** or **Gen Y** (1977–1995), existence is less about status and prestige, but rather a healthy work–life balance and the "here and now". There is a greater willingness to be mobile, to travel, to discover. Generation "Why" does not assume anything, but questions established rules and expectations, less out of belligerence than curiosity: if it makes sense, they will do it. This is a generation that combines work with life.
- **Centennials**, **Gen Z** or **iGen** (post-1996), challenge in turn the values of their predecessors. Given the opportunities opened up by people and technological connectivity, this is the most diverse, internationally versatile, technological savvy and culturally immersed group so far. Everything is posted, shared, commented upon. For this generation, work is just one part of life.

Each of these generation groups has a different expectation of how the working environment needs to be shaped, which can create a see-saw of expectations between employee and employer. Finding that balance is one of the key success factors in creating diverse and inclusive working environments. This chapter will focus initially on some indicative workplace trends, plus a brief look at France's workplace environment, and will then consider three key areas:

1 the multigenerational workplace;
2 an inclusive workplace for those with disabilities;
3 the family-friendly workplace.

4.2 Indicative trends

In early 2017 RICS conducted a *Europe Chairs' Insight Survey* of the senior representatives of its national boards in 13 European countries. The aim was to get an indicative (rather than representative) view of attitudes towards D&I given the cultural differences throughout Europe, focusing on two particular areas: diversity and business practices, and inclusion and cultural barriers.

Question 1: Diversity and business practices

Think about how employment diversity is dealt with in your business environment. Which of the following areas/events should be prioritised? Please rank by order of priority (from 5= high to 1= low).

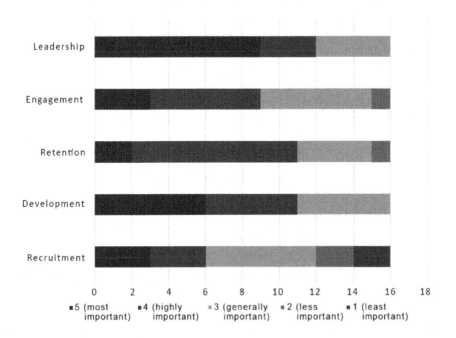

Figure 4.1 Diversity and business practices

Responses indicated that while D&I is on the agenda, there was a clear statement that change needs to start at the top, with leaders and senior management actively and visibly supporting the case for D&I through words and deeds. Collective responses of most/highly important verify that developing and retaining staff is a key priority, but integrating D&I into recruitment practice still needs to be given more attention.

Question 2: Inclusion and cultural barriers

In your opinion, which of these groups do not have equal chances and opportunities at work?

Responses indicated that people with disabilities are in the most vulnerable position when it comes to D&I being integrated into the organisational culture. Ranking at second and third were people with a different ethnic background and senior employees, closely followed by women.

Survey responses confirmed that in some countries the legal framework also needs to change. Some countries are already well advanced, while others continue to face barriers (see also Chapter 1, section 1.3: Legislation). This is not a dissimilar pattern to any new trend or development – what is important is that the trend has started and is irreversible. People, often due to religion or cultural background, have difficulties in accepting that others may be "different". There are always early adopters, and there will always be the late majority. The good work needs to continue in order to change this over time and show not just the monetary, but also the societal perspective of embracing a diverse and inclusive culture. On the one hand, people need to take their time to adopt to

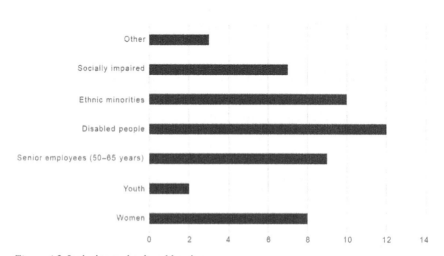

Figure 4.2 Inclusion and cultural barriers

new realities, and on the other hand, it is also the responsibility of those who support the case of D&I to persevere in enlightening and challenging those who still need to be convinced.

4.3 Understanding cultural differences: France

To try to unpack just some of the cultural considerations in considering D&I in the workplace, we turn our attention to a country – France – which has introduced some leading-edge legislation and which is therefore a good example to learn more about attitudes to D&I and what is being done there to attract talent to the industry.

France, as in many other markets, is faced with the lack of skilled professionals in an increasingly competitive sector and an increasingly global world, constantly evolving and changing in the face of disruptive trends such as climate change, new technologies and demographic changes. In discussion with senior real estate professionals in France we have learnt that it is crucial for companies to adapt and attract new talent to the industry and reinvent their USP offering in order to retain the changing workforce. Challenges above and beyond this, and measures which are being put in place to address them, include the following:

1 The number of complaints related to discrimination at work is high in France, and in particular with regard to "inconvenient" questions during the recruitment interview about age or gender (e.g. maternity or pregnancy). Yet criminal and labour law is clear that these are discriminatory recruitment practices.
2 Fortunately, over the past ten years new laws have been introduced in order to reinforce stricter measures in relation to diversity, and as a result, companies which are found to be non-compliant risk being penalised. There is no adverse discussion about the role and added value of women in the profession since commercial law prescribes gender-balanced representation on a board.
3 National legislation encourages the employment of people with disabilities. Public and private firms with more than 20 employees are required to employ at least 6% of disabled people,[1] or pay an annual contribution to l'Agefiph,[2] the association in charge of facilitating the inclusion of people with disabilities. Furthermore, companies with 50 employees and more are required by law to have an action plan in place in order to encourage employment of older people and women, for example.[3]
4 Some organisations, and not only those working in the land, construction and property sector, have gone a step further in their commitment and signed the French Diversity Charter,[4] which focuses on raising awareness of the charter among SMEs and micro-enterprises and which is interlinked with other European charters (see also Chapter 1, section 1.4: International firms versus SMEs).
5 The private sector can also play a role in influencing public policies to develop sustainable neighbourhoods. The cost of failing to support young, qualified, but socially disadvantaged professionals is high compared to the small investment necessary to facilitate induction, coaching and training

to help them get the jobs they deserve. An example of this is Nos Quartiers ont des Talents,[5] a French association that facilitates business networks and employment commensurate with qualifications independent of socio-economic background.

From a recruitment perspective, there is no policy of quotas to identify the right profile demanded by employers. However, trends are changing in the recruitment of real estate professionals, mainly driven by the way globalisation is impacting the property sector and the way it functions.

As in other countries, there is still a big difference between local or small companies and larger international firms when it comes to openness and flexibility. While smaller firms remain interested in traditional profiles, larger ones are looking for new professionals who are not the conventional type and who have a broad range of competencies and experience, such as social skills and ease with new technologies.

4.4 The multigenerational workplace

Managing and aligning generational diversity is a challenge to the corporate culture. Sinking birth rates on the one hand and an increase in the average working life on the other mean that having three, if not four, generations in the workplace is becoming the norm. Each individual generation has been conditioned and shaped by the social, economic and political environment of its own particular time.

Looking beyond the present and using the variety of skills as an advantage to foster cross-generational acceptance and collaboration benefits the business. Each generation has its own characteristics that may be applauded, such as the digital versatility of those from Gen Z, or even snubbed, such as the negative connotations of ageism frequently associated with the Baby Boomer generation. Ageism joins the long list of prejudices, as it holds the belief that older people have less value than younger people, not just in a social context, but also in the workplace. The many benefits of attracting "the next generation" are dealt with elsewhere in this book; therefore, to give a balanced perspective, the next section will focus on the "older employee".

How "older employee" is defined is usually a subtle mix of conscious and sub-conscious factors: age (assumed), experience (factual or assumed) and appearance (subjective). In any event, the description that somebody has "years of experience" tends to sound more positive than "has been in the organisation for years". Research by Catalyst in 2018 explained in more detail how generational differences are not just about age, but a complex combination of circumstances as well as number of working years and experience, and that these in turn are all punctuated by expectations and needs: "when it comes to managing workplace diversity age, per se, is not one-dimensional".[6]

A number of years ago an HR business partner was engaged in conversation with two employees. She made the claim that the average length of service in an organisation should be around six to eight years, and after that time people should move on, leave the company and change jobs. It was not clear whether the HR business partner was aware that both employees, one a Baby Boomer and the other Gen X, had been in the organisation for more than 15 years. The assertion was both thought-provoking and disrespectful, and is an example of why age and length of service earn their places alongside other sensitive issues. Let us consider two scenarios.

Company X provides little opportunity for staff development, and as a result, Employee X is most likely mechanically and repetitiously doing the same work in the same way day in and day out. This does not necessarily mean that they are less committed to their work or less loyal to the company, but there is a higher risk of innovation and creativity becoming stifled along the way.

Company Y, on the other hand, encourages staff to develop new skills and learn a language relevant to their work, and provides secondment opportunities, travel opportunities to connect with peers from different departments in different locations, etc. What emerges is a different job awareness and open-mindedness. The 15 year-plus employee remains alert to wider organisational issues, adapts, bonds, teamworks with others, and adds additional value through long-standing historic knowledge and experiences.

It is therefore imperative for the organisation to offer opportunities to all employees, irrespective of age, experience or generational "labelling". The following examples present some ways to bring out the best in "older, experienced or long-term" staff:

1   **Provide mixed-generational networking opportunities that break down homogeneity and stereotyping.** Get people talking to and understanding each other. This can be achieved through something as simple as an informal session during a lunch break, or a more formalised interaction as part of a mixed-skills project management group.

2   **Develop a programme of flexible working and shared parental leave, and recognise those with caring responsibilities.** While parental leave and childcare may be a priority for younger workers, older employees are frequently challenged with caring for sick or elderly parents and relatives. A commitment to acknowledging this and giving equal consideration in recruitment and retention policies will help attract and retain employees of different age groups.

3   **Encourage a mentoring and reverse mentoring programme across the organisation.** Faced with increasing performance- and target-related expectations, employees of all ages and experience are challenged to acquire new competencies. While Baby Boomers and Gen X may be lacking on social media skills, they most likely have a better portfolio of leadership, resource management or decision-making capabilities – skills which Gen Y and Gen Z may only have periphery experience of. While "learning by doing" is still a key maxim, there is at times no margin for error in making some decisions, therefore additional support needs to be put in place. This is where mentoring can help.

Mentoring and reverse mentoring are an effective means of growing and developing people of all ages. While separate from formal line management, (reverse) mentoring can be one element of an internal diversity strategy, or employees can be given financial support to participate in external programmes. An effective mentoring programme experienced first-hand has been IMMOMENT,[7] the first cross-generational mentoring programme launched in Germany in 2015 and now in its third year. Supported by RICS, 20 senior real estate managers from leading organisations, such as Arcadis Deutschland, BerlinHyp AG, CBRE Real Estate, Corpus Sireo Real Estate and the German Property Foundation (ZIA), have volunteered as partners to 20 "matched" juniors to support their career aspirations. The key themes that have emerged are:

- **character development** – e.g. help to become more resilient and agile, and keep pace with changing work environments;
- **value-definition** – understanding the meaning of professional ethics and defining personal values;
- **tackling a specific challenge or (perceived) weakness**;
- **helping mentees to formulate a career vision** – "Where do I want to be in x years' time?";
- **learning to take the initiative and identify new opportunities** within an existing organisation – this can be a linear progression or a sidestep into a different department or team in order to develop a portfolio of projects and experiences.

As with all mentor–mentee partnerships, the above relationships had no pre-set agenda, and it was only during the course of the relationship that decisions were made relevant to "making a change".

Another aspect dividing multigenerational understanding is digital disruption. The explosion of social media, intra-company chat sites and WhatsApp messaging have all been contributing factors to how communication styles are changing in the business environment. A frequent lament from the older/experienced/Baby Boomers is that formal language and tone are rapidly being replaced by (too) informal and chatty writing styles, coupled with the fact that meaning is being conveyed increasingly by expressing intent through pictures, or emoticons, rather than words, and that discipline and accuracy are being crushed by sloppy punctuation. Are badly formulated reports a sign of disrespect, or a sign that employees are losing the ability or discipline to write long reports as they are expected to cram as much as possible into the day in order to get everything done? For an organisation to be fit for the future, creating clear communication style policies can establish norms and expectations to be followed by all employees irrespective of age.

4.5 An inclusive workplace for those with disabilities

By far the biggest discrimination in the workplace is in terms of disability, with people with disabilities being three times less likely to be in work.[8] At the heart of the issue are employers who do not want to either adapt the physical workplace to accommodate those with disabilities or provide the additional support that may be required to enable a disabled person to perform at their best in the workplace. It is interesting to note that based on current statistics, only 0.6% of RICS professionals are disabled. Yet according to the World Health Organisation and the World Bank, about 15% of the world's population live with some form of disability, of whom 2–4% experience significant difficulties in functioning. That is around 1 billion people globally. Improvements in measuring and recording those with disabilities plus, of course, an ageing population have all contributed to the increase in this figure from around 10% in the 1970s.

But what does this mean for real estate and construction? Firstly, we need to understand the form of disability, which can be physical, mental, or both. Plus we need to bust some myths – it is not correct that people with disabilities cannot work in real estate or construction. But we do need to understand and make allowances for those with disabilities in the workplace. For managers who work with and manage people with physical and mental disabilities, understanding is the first imperative, followed by adaptation.

Disability is acknowledged as one of the "protected characteristics" identified as part of D&I in the workplace. In order to optimise talent in the sector, it is important for organisations as employers to better accommodate those with disabilities in order to recruit and retain all of the best talent available.

Talking about having a disability is no longer frowned upon as societal change has become more accepting of those with disabilities. Physical disabilities, mental illness or an illness that prevents a normal life are all forms of disability. Some are, of course, easier to spot than others, but encouraging an environment where employees feel safe to disclose and discuss their disability with their manager/employer is a very important first step. There should be no stigma in discussing disability at work, and creating an environment and culture that allows people to speak freely about a disability ensures both parties get the best from each other in the workplace.

Often legislation in countries, such as the UK Equality Act 2010,[9] drives social policy change, but in some cases it can also set the bar too low to encourage pushing the boundaries on more openly challenging social norms and embracing disability within the workplace.

Supported by legislation from governments, the workplace is becoming more readily adaptive to those needing changes, and travelling to or from work is made easier by initiatives such as "step-free access" or allocated parking spaces for those with disabilities. It is imperative that we try to better understand disabilities because no two people with a disability are likely to be the same, therefore organisations need to plan for accommodating people with a disability into the workplace alongside other colleagues. Sharing ideas and understanding is all part of the D&I journey, so helping all staff to embrace enlightened thinking and understandings will help provide a better and more inclusive environment for everyone such that everyone feels able to bring themselves to work every day.

From personal experience as managers, people with disabilities need to feel unhindered in talking about their disability with their manager as the representative of their employer so that their needs can be understood and accommodated without fear that by disclosing their disability, they will either be adversely affected or treated differently.

But on some things employers do need to think differently to help those with disabilities feel comfortable at work. Some examples of adapting the workplace to positively embrace those with disabilities include:

1   people with disabilities being provided with clear information, in a format that they can understand, about what support or changes to the workplace can be made available to them;
2   considering how buildings, services and the workplace environment can be adapted so that they can access and use it safely and freely, alongside their colleagues;
3   thinking about what additional training and support may be required to help them do their job effectively in the work environment they choose;
4   providing help and support before things get on top of them – this may mean that managers or colleagues need to be supported to understand what signs to look for, and then how to respond and find effective solutions to deal with the situation;
5   thinking about what technology and gadgets can be provided to help support people to be more independent and to work more effectively;
6   enabling flexible working that allows people to not have to always be in the office if they need to work closer to home, or at home, from time to time;
7   providing support to keep their job if they become disabled while they are employed.

Whatever support is provided, it should of course be tailored to the individual's needs. Helping colleagues at work understand the needs of a person with disabilities and providing guidance to managers to better support their teams are all part of creating an environment that is not only diverse and inclusive, but positively embraces people with disabilities in the workplace so that everyone can perform to the best of their ability.

4.6 The family-friendly workplace

Global research by EY in 2015 involving around 9,700 full-time employees in the US, UK, India, Japan, China, Germany, Mexico and Brazil concluded that it is becoming increasingly difficult to manage a healthy work–life balance, with younger generations and parents hit hardest. After competitive pay and benefits, one of the top priorities scored by survey respondents was "being able to work flexibly and still be on track for promotion".[10]

From discussions with real estate professionals, it is clear that the sector offers exciting and challenging career and development opportunities, but there are also barriers to progression, and frequent complaints about the sector being too conservative to successfully facilitate the balance between home/family and work.

Women in particular often get caught in the "part-time trap", and in real-life situations they frequently experience being overlooked for promotion despite being able to consistently demonstrate good results.

*"Our mom does the same amount of work as everyone else, and we do timesheets. Promote us!
I mean, promote her."*

Parental challenges are often at the heart of discussion, but there are also challenges around caring for sick or elderly relatives over the short or long term, and these may be exacerbated by increased work mobility meaning that the child or sick relative may be in one location or country and the carer in another.

Being a parent and holding down work commitments is not easy for anyone, at any level. There are always conflicting demands on your time. Consequently, it is an all too familiar scenario that colleagues who have small children of pre-school or junior school age have to juggle the family agenda each week: an early rise, get the children ready for their day, a mad dash to drop them off at childcare or school, and in the afternoon or evening, take care of activities, homework and meals, or just "be there". It would help encourage diverse teams if organisations could allow each family to have the right to determine what works best for it, noting that one of the most negative experiences is for another person to judge how your family is balancing childcare with work and passing judgement on whether it's enough or not enough. The psychological pressure of not getting home on time when a child is waiting for you, knowing they either have to let themselves in with a key or wait on the doorstep, is enormous. Of course, this is not the employer's responsibility, but what is important is a level of understanding and an ability to discuss acceptable working arrangements, so that employees can make suitable and reliable childcare arrangements.

In order to delve deeper into the challenges of balancing a young family with work, RICS invited female real estate professionals from international and locally based organisations to a series of round tables in order to debate the challenges they have faced personally and to identify any collective recommendations or solutions. The five round tables, titled "Women in Real Estate: An Asset for the Sector", were held in five major cities in Germany (Munich, Frankfurt, Berlin, Cologne and Hamburg) during 2016–2018 with exclusive invitations to over 50 female professionals.

What stood out most was the vigour of the discussion resulting from the pent-up emotions of balancing work with childcare. Once the door was closed and the participants were in a safe and confidential environment, it was almost impossible to stop the stream of words and comments. It was as if the participants needed a conduit for frustrations about raised expectations and the realities of the situation – for example, frustration that there is no consistent management approach to gender parity, that it is often used as a marketing tool to attract talent to the company, and that career progression can be severely restricted depending on the employer.

Topping the insight list was **flexibility** in terms of the following:

1   recognising the importance of being able to see flexible working role models (both women and men) to emphasise that a healthy work–life balance is equally attractive to both women *and* men;
2   facilitating the challenge of caring for young families as well as parents and relatives needing care and attention;
3   flexible working being seen as a decisive element for organisations to increase their competitive edge, recruit high-potential employees, retain qualified staff and maintain high engagement levels;
4   traditional working models are too focused on being physically present, yet this does not necessarily guarantee high engagement;
5   result-based working models, irrespective of a home or office location, can still be measurable on the basis that it is not *where you are* but *what you do* that counts. This needs to be fact-based and completely transparent.

But the participants not only expressed frustrations, but also many insights and ideas. Observations and conclusions from discussion groups like the above lead to recommendations whose impact each organisation can assess, then consciously develop and embed policies and culture to facilitate parental and care working for those who may need it most, while at the same time offering flexible work-ing within agreed parameters for all employees, within the limits of their role requirements, and communicate this across the organisation so that the policy is equal, fair and transparent to all. Creating the balance between child/parental/ relative care requirements and transparent policies which do not in turn put those without such needs at a (psychological) disadvantage is one of the challenges a CEO needs to address.

4.7 Message to the CEO

As the CEO, you are continually under pressure to find technically competent staff with the right level of skills, experience and capabilities. As lifespans increase, a multigenerational workplace of up to four generations is becoming the norm rather than the exception.

Developing the right talent pool is complex, and to attract new talent, more emphasis needs to be placed on workplace considerations. The employee mix ideally should contain people from different geographical and social backgrounds, as well as from different cultures and with different value systems. Plus, of course, you need to balance these with the context of your organisation and the fact that the different skill sets offered by each generation have to complement and balance the workforce as a whole. Then there is the fact that gender parity needs to be encouraged and supported at all organisational levels, cascading down from the top (leadership role models) to the lower organisational levels.

"These employees look diverse and delicious—I'll take one of each."

Central to creating the ideal workplace environment is the ability to:

1   look beyond the generation (age) or number of years of service, and look instead at the overall contribution;
2   develop human resources policies that are family-friendly;
3   have faith in those employees who are working flexibly away from the office;

4   introduce a results-based working philosophy, rather than relying solely on counting on a physical presence in the office;
5   develop career pathways that encourage lateral changes as well as upward progression, to help build a portfolio and adaptive career pathway.

Your ability as an employer to be agile and flexible is crucial for the selection, recruitment, motivation and long-term contractual commitment of your employees.

Should you still be at the start of the D&I journey, be open-minded about best practice examples from other organisations.

Providing training and mentoring on D&I will help your managers and employees to increase their diversity-awareness, eliminate unconscious bias, and encourage them to think differently and more openly about inclusivity. Human resources departments need to scope training and guidance that is rolled out to all hierarchies of the organisation, starting with the C-suite directors, right through senior and middle management down to departmental and team levels.

Notes

1   AtouSanté (2018). The French law on disabled persons. Available from: www.atous ante.com/en/french-law-disabled-persons/ [Accessed 30 September 2018].
2   Agefiph (2018). Available from: www.agefiph.fr/ [Accessed 30 September 2018].
3   Employment and Labor Lawyers International (2014). New French Works Council legislation impacting business with 50+ employees. Available from: http://ellint.net/news/sector/international-secondments-business-immigration/new-french-works-council-legislation-impacting-business-50-employees/ [Accessed 30 September 2018].
4   Charte de la Diversité (2018). Available from: www.charte-diversite.com/ [Accessed 30 September 2018].
5   Nos Quartiers ont des Talents (2018). Available from: www.nqt.fr/ [Accessed 30 September 2018].
6   Catalyst (2018). Generations: demographic trends in population and workforce. Available from: www.catalyst.org/knowledge/generations-demographic-trends-population-and-workforce [Accessed 30 September 2018].
7   IMMOEBS e.V. (2018). Mentoring-Programme IMMOMENT. Available from: www.my-immoebs.de/cms.550/show/dasmentoringprogramm [Accessed 30 September 2018].
8   Stats NZ/Tatauranga Aotearoa (19 August 2018). Disabled people three times less likely to be in work. Available from: www.stats.govt.nz/news/disabled-people-three-times-less-likely-to-be-in-work [Accessed 30 September 2018].
9   legislation.gov.uk (2010). Equality Act 2010. Available from: www.legislation.gov.uk/ukpga/2010/15/contents [Accessed 30 September 2018].
10   EY (2015). *Global Generations: A Global Study on Work-life Challenges across Generations. Detailed Findings*. Available from: www.ey.com/Publication/vwLUAssets/EY-global-generations-a-global-study-on-work-life-challenges-across-generations/$FILE/EY-global-generations-a-global-study-on-work-life-challenges-across-generations.pdf [Accessed 30 September 2018].

5 The principles of building an inclusive workplace

5.1 Introduction

This chapter is primarily based on the learning outcomes following the roll-out of the IEQM (Inclusive Employer Quality Mark) launched by RICS in 2015. The mark was developed for use within the UK market, but the principles are globally relevant and can be adapted depending on local law and cultural sensitivities around the world.

Following the launch of the IEQM, data was collected and a research project undertaken involving firms, small and large, from across the UK aimed at forming a baseline for employers to come together to provide a level of commitment and to share leading practice. A study conducted by EY in 2016 on behalf of RICS brought together the key findings from the analysis of the data provided by the employers, providing the first baseline of the sector by the sector. This culminated in a report called *Building Inclusivity: Laying the Foundations for the Future.*[1]

While the IEQM is being updated by RICS and reduced from the original six key themes to the four of leadership, recruitment, culture and development, in our book we will use the original six principles outlined in the report by EY as a framework for understanding the issues for real estate and construction that arose from the report. The principles remain robust, and the findings here are relevant in helping any organisation looking at D&I in the workplace, providing a framework against which they can start to formalise their own approach to D&I. The additional principle RICS is introducing on culture was covered earlier, in Chapter 4.

5.2 The IEQM benchmark

In order to make a real difference to the sector, a principles-based approach was initiated to help organisations that employ property professionals to adopt leading practices for diversity and inclusion in the workplace.

The IEQM aims to drive behavioural change by encouraging all organisations, large and small, to look carefully at their employment practices and to put diversity

"INKlusion and DiversiTEA are at the HEART of IEQM!

"Dude—is that a real human heart?"

and inclusion at the heart of what they do. This voluntary IEQM standard is designed to help organisations in the profession gain a competitive advantage as well as develop a more diverse and inclusive workforce. It provides a simple self-assessment process any organisation can use, and any organisation participating in the process is allowed to use the IEQM mark as a sign of its pledge to work towards the D&I agenda.

The IEQM asks employers to pledge their commitment to adopting and continually improving against six globally applicable principles that signal to the outside world that the organisation is serious about diversity and inclusion in the workplace.

It is worth noting that the IEQM is not designed to be a quality assessment of standards or competency, and is focused on organisations within the RICS collaboration network. This is in contrast to other D&I standards, such as the National Equality Standard (NES)[2] that provides businesses with a recognised national standard in the world of diversity and inclusion for them to reach and then build upon through independent assessment.

There are, of course, several other types of assessment available globally, but the IEQM principles and learnings are a solid place from which to start to build an organisational D&I vision and policy. The six principles established as part of the IEQM focus on:

1   **leadership and vision** – a demonstrable commitment at the highest level to increasing the diversity of the workforce;
2   **recruitment** – engage and attract new people to the industry from under-represented groups using best-practice recruitment methods;
3   **staff development** – training and promotion policies that offer equal access to career progression for all members of the workforce;
4   **staff retention** – flexible working arrangements and adaptive working practices that provide opportunities for all to perform at their highest levels;
5   **staff engagement** – an inclusive culture where all staff engage with developing, delivering, monitoring and assessing the diversity and inclusivity policies;
6   **continuous improvement** – continually refreshing and renewing the organisation's commitment to being the best employer, sharing and learning from best practice across the industry.

Each principle has a series of proof points (criteria) that show member organisations what leading practice and success look like for this sector, and form the framework of their self-assessment. By submitting their data, participating

Figure 5.1 RICS Inclusive Employer Quality Mark

organisations receive an individual benchmark analysis finding which allows them to understand shortcomings and take action for improvement. Firms that have participated in the IEQM may also display the IEQM logo to show they are committed to this agenda, as well as becoming part of the Inclusive Employer network that gives them exclusive access to a database of leading practice examples.

By 2018, some 177 organisations had signed up to the IEQM across the UK, representing over 300,000 people.

At the time of writing, and following feedback provided by professionals and firms across the sector, a review of the IEQM is under way by RICS to ensure that it remains fit for purpose and provides a worthwhile quality mark that carries both esteem and value for signatories. There are no current plans to extend this outside the UK at present, but the principles can, of course, be applied more widely.

One of the areas also under review is the nature of the self-assessment survey. With 190 questions, many small firms in particular highlighted lack of resources as being a key reason why they could not complete the survey. The decision was taken that the IEQM will in future be based around four, rather than six, core principles:

1   **leadership** – a demonstrable commitment at the highest level to increasing the diversity of the workforce;
2   **recruitment** – engage and attract new people to the industry from under-represented groups using best-practice recruitment methods;
3   **culture** – an inclusive culture where all staff engage with developing, delivering, monitoring and assessing diversity and inclusivity;

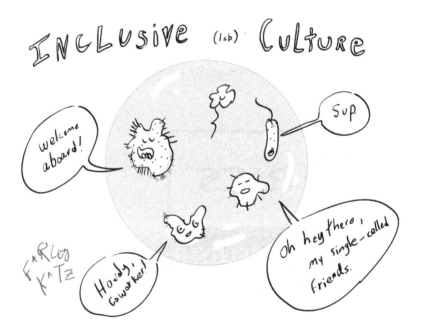

4 **development** – training and promotion policies that offer equal access to career progression to all members of the workforce.

Barry Cullen, Diversity and Inclusion Director for RICS in the UK, explains the rationale and aspiration behind the changes:

> This will enable a revision of the self-assessment, a reduction in the number of questions, and a tighter set of principles in line with current issues and challenges.
>
> Key to this is the inclusion of **culture** as a principle. Real estate and construction have been sectors where a non-inclusive culture has been prevalent. RICS is committed to raising the awareness of surveying as a career choice with the ambition for the profession to be more diverse, but this is only possible if a more inclusive culture exists. There must be a measurement of this to determine how inclusive organisations are.
>
> With a tighter, more accessible model, the IEQM will appeal to a broader audience of organisations. By working with UK regional teams, there will be engagement with smaller businesses to help with their understanding of the value of the IEQM and the benefits of a more diverse and inclusive business. Through this work RICS will be at the forefront of the sector ensuring a dynamic, diverse and inclusive profession fit for the 21st Century.

5.3 Leadership and vision

Demonstrable commitment at the highest level to increasing the diversity of the workforce is essential. This commitment must start at board level within the organisation, and requires the commitment of the CEO and the entire team to be effective.

"Now that's some diverse DiversiTEA."

Employees will see through passive commitment, and it is imperative that each and every day each member of the senior team literally "walks the talk" when it comes to embracing the principles and values of the organisation associated with D&I. Committed leaders with a clear vision give credibility to the cause, and set the tone for the rest of the organisation.

"What—you've never seen somebody walk the talk?"

It is important that a clear vision for what the D&I agenda means for the organisation and how it is going to be taken forward are clearly documented. This needs to involve the human resources team, but must be seen to come from the wider leadership as a whole in order to be effective. Setting a policy for D&I in the organisation with clear, measurable, goals is an important first step.

Once the vision has been agreed at board level, it is important to establish a D&I impact assessment (or similar review mechanism) attached to leadership decisions, so that the business can evaluate the positive or negative effects any initiative or activity has on diverse groups. D&I only truly becomes embedded when it is considered as part of all business decisions, and is part of the overall business strategy and organisational ethos. Leadership D&I training will help leaders within the organisation fully understand and embrace the D&I agenda, as well as provide them with the tools they will need to be successful. It is important that those who do not uphold the values of the organisation are brought into line.

One of the key messages from the research is the importance of role models and mentoring in helping support the leadership agenda. The CEO, board members and executives are seen by their employees as role models within the organisation. Role models are an essential part of leadership and vision, and encouraging staff to embrace role models, at all levels, is unquestionably a way of embedding and encouraging engagement outside the normal line management responsibilities. As role models, it is important to instil in the leadership team the values and behaviours that are key to the organisational culture and success.

"My role model is a French Press."

"When I grow up, I want to be extruded from a fountain pen."

Mentoring and reverse mentoring are great ways to enable people outside the formal direct management lines to connect and learn from each other. Reverse mentoring is becoming increasingly popular in connecting senior people with more junior people in the organisation. This is where a perceived more junior person mentors a more senior person to help provide a different perspective and a direct conduit through an unofficial channel. This can also be an effective way to check that the leadership vision and values are working throughout the organisation.

Here are some suggestions to take into account:

- Clear role models, with whom individuals from diverse backgrounds can identify, can inspire such individuals to emulate the successes of their role models.
- Where internal, diverse role models do not currently exist, organisations have looked at inviting external role models to speak about their experiences.
- Role modelling is not just about individuals' backgrounds – it is also about inclusive behaviour.
- Leaders who exude inclusivity through their management style and consideration of individuals' needs can be powerful role models.
- Lack of visible role models is cited as a reason by some job candidates for not engaging with the industry.

Key observations from the research project

- The D&I agenda and vision need to start at the most senior level within the organisation in order to be successful.
- Having a D&I policy is an important first step, as it is a public commitment to the agenda through initiatives like the IEQM (others include the National Equality Standard).
- Leadership involvement, through action plans, is essential in monitoring the development and retention of a diverse workforce.
- It is important to have a D&I impact assessment (or similar review mechanism) attached to leadership decisions, so that the organisation can evaluate the positive or negative effects any initiative or activity has on diverse groups.
- D&I becomes embedded only when it is considered in all business decisions, not just limited to HR decisions, and is part of the overall business strategy.
- Leadership D&I training will help leaders fully understand and embrace the D&I agenda.
- Having strong role models and using mentoring within the organisation really helps embed and instil the D&I culture.

5.4 Recruitment

Recruitment is an important process in the growth and development of any organisation. Recruiting people who will fit within the organisational culture and will be successful is critical. Recruitment can be expensive and time-consuming, and is an important process where D&I needs to be considered and embraced by any organisations that are serious about embedding D&I.

Practices often lead to recruitment of those who are similar in background to the cultural DNA or to the recruiter – a form of cloning. It is important to avoid occupational segregation that reinforces the idea of jobs being done by a particular employee type, particularly in terms of gender, as this restricts the pool of talent for recruitment purposes. Putting practices and processes in place to remove aspects of unconscious bias and thereby allowing for an open and informed process that embraces D&I is an important step in making and embedding organisational change towards a more diverse and inclusive organisation. It is important to engage and attract new people to the industry from under-represented groups, and to gather and share best-practice recruitment methods. Inclusive recruitment processes and programmes help to build and nurture the diversity of an organisation's talent pipeline. Looking at broader recruitment policies, such as an internship or apprenticeship programme, can help broaden the inclusivity pool. Equally, looking at a "returners" programme to help attract leavers back into the organisation, particularly mothers, is an effective way to build greater diversity. Initiatives to bring greater diversity into the sector should be supported and encouraged, particularly when bringing in mature or non-cognate (for example, graduates without a RICS-accredited or real estate degree) entrants from outside the sector – from the armed forces or the banking sector, for example.

"I'm cloning new workers!"　　　*"DiversiTEA can NOT get behind this."*

With diversity, such as gender parity on boards, being more widely scrutinised, with it becoming a legal requirement in the UK to publish the gender pay gap, and with Iceland leading the way on gender pay, it is vital that recruitment considers all aspects of D&I, as this is the key entry point for the organisation and establishes the organisational pyramid (see Chapter 3).

In addition, gender pay equality is a legal requirement in many countries, and this is certainly a key consideration many future employees, especially graduates, look for when selecting a company they would like to work for, and is considered an absolute essential for any organisation seeking to attract top talent.

Key observations from the research project

- It is important to adopt leading practice recruitment methods.
- Consider D&I throughout the recruitment process.

(continued)

(continued)

- Consider engaging and attracting new people to the industry from under-represented groups.
- Starting salary and parity across male and female recruits is an important consideration.
- Apprenticeships and internships are effective ways of bringing new talent into the organisation.
- Consider recruiting non-cognate or mature entry candidates or those from outside the sector, such as from the armed forces, to diversify the intake.

5.5 Staff development

Once people have been recruited and become employees, staff development and subsequently retention become core components of the organisation's D&I commitment. The organisation should encourage everyone to feel able to "bring themselves to work every day" so that no matter who that person is, they feel they are being treated fairly and equitably alongside peers.

"I'm bringing myself to work today."

Developing people through training, mentoring and coaching is an important part of growing and nurturing organisational talent and helping to support people through their careers within the organisation. Equal and transparent training and promotion prospects enabling all staff to develop and succeed are an important aspect of the D&I culture. Fair access to training and appraisals helps to ensure that organisations are developing and progressing their best talent. It is therefore important that training is inclusive. So the next time you are organising training, consider timings, location and delivery methods and ask the simple question of whether this is the optimum time, place and approach that will allow everyone to come and participate.

Key observations from the research project

- Equal and transparent training and promotion prospects enabling all staff to develop and succeed are important aspects of the D&I culture.
- Fair access to training and appraisals helps to ensure that organisations are developing and progressing their best talent.
- Pathways for requesting training must be transparent to all employees.
- It is important to take into account the accessibility of training locations, timing and delivery methods.
- Appraisals are essential, as is monitoring an individual's development through a personal development plan so that impact on D&I can be better understood.
- Training managers about unconscious bias to reduce bias in appraisals and promotion choices is essential.
- D&I goals should be considered when implementing fast-track leadership programmes.

5.6 Staff retention

Recruiting the right people, developing and training them are all essential aspects for an organisation that is based on its people providing a service. Retaining your best staff is a business imperative, as they need to become the future of the organisation and are tomorrow's leaders. Losing good people is not what an organisation wants, no matter how effective its alumnae networks may be.

Keeping the people you want in the organisation so that they are achieving and delivering their best is an inherent part of any people-based business. Flexible working arrangements, adaptive working and wellbeing practices, and pay equality all help in making yours a workplace of choice. The ability to retain employees is crucial for today's organisations, who all compete for talent.

"Please—there are easier ways to retain your best staff."

With this in mind, having a plan to address any unequal pay issues is essential. Equally, supporting staff during periods of absence (longer than two months) and helping them back into work are recognised leading practices. Monitoring implementation of return-to-work policies needs to become the norm.

Encouraging support networks enabling employees to come together with other like-minded people is another great approach to help employees support one another at work and to, hopefully, provide key support when required that may also help avoid unwanted attrition. Some excellent examples include multi-faith networks, women's networks, LGBT networks, minorities networks and parental (or mother) networks. Although it is acknowledged that not all of these will be acceptable in every country, encouraging groups of staff to meet informally and support each other helps build a stronger organisational cultural DNA, as well as embed a sense of values and belonging, all of which are essential for staff retention.

Key observations from the research project

- Flexible working arrangements, adaptive working and wellbeing practices, and pay equality all help in making yours a workplace of choice.
- Action plans to address significant gender pay differentials are essential.

- Flexible benefits packages to encourage retention of staff with diverse needs need to be considered – these could include value and recognition awards, for example.
- It is a good policy to consider adaptive meeting times and formats to take account of different working practices.
- Supporting staff during periods of absence (greater than two months) and supporting them back into work are recognised leading practices.
- Monitoring the implementation of return-to-work policies needs to become the norm.
- Increasing adoption and uptake of formal flexible working is likely to become the norm, not the exception – supporting staff and managers will be an important step in making this successful for the organisation.

5.7 Staff engagement

Staff engagement is about everyone within the organisation feeling empowered to be part of it and to be involved in the D&I agenda – an inclusive culture where all staff engage with developing, delivering, monitoring and assessing the D&I of their workplace. To optimise engagement in organisations, here are some tools to consider.

"I'm engaged! I'm empowered! I'm about to become the world's limberest surveyor!"

Employee surveys

Employee engagement is about two-way communication and feedback from the top of the organisation to the bottom, and vice versa. Employee surveys are a good way of anonymising and capturing staff sentiment, and employee engagement surveys provide an increasingly popular and useful tool to achieve this while also providing an opportunity for staff to offer feedback.

Employee groups

Another way of providing feedback from employees is through employee groups. Quite often, bringing employees together from different parts and layers of the organisation leads to unexpected but relevant ideas to improve the organisation. Having said that, employee groups can only be successful if they are taken seriously and proposals are discussed at senior/board level.

D&I champions

D&I champions help to drive the agenda forward, as they take personal responsibility and interest in doing so. They also help to raise awareness and can articulate the issues in a meaningful way at a local level, and in this way they "mainstream" the D&I agenda.

Building D&I awareness

Aside from unconscious bias training and inclusive leadership training, organisations also reported providing D&I training on areas such as:

- cultural awareness;
- disability and mental ill-health awareness;
- wellbeing.

Training staff in these areas provides them with understanding, skills and the language needed to help change the culture of their organisation and make it more inclusive.

Key observations from the research project

- Employee feedback is essential, and the consistent use of staff surveys or employee groups to gain employee feedback are seen as effective mechanisms in common practice.
- Having D&I champions to drive change at a local level works well when coupled with senior-level engagement.
- Building D&I awareness and training for all staff provides them with an understanding of the skills and the language needed to help change the culture of their organisation and make it more inclusive.

5.8 Continuous improvement

Continuous improvement involves continually refreshing and renewing the organisation's commitment to being the best employer, sharing and learning from best practice across the industry. Reflecting on feedback, acting on action plan monitoring and driving organisational change are all part of the continuous striving for improvement in the D&I agenda.

"Don't you just LOVE reflecting ... on feedback?"

An organisation does not remain static in itself or its sector, with external market, political, economic and social-economic trends continually affecting the environment. An organisation needs to be nimble and adaptive to the changes around it with regard to diversity and inclusion.

Regardless of the level of D&I maturity of an organisation, there are always opportunities to improve and set more challenging goals. Reviewing D&I activity allows the organisation to reassess what it should start, stop or continue doing. Moreover, review and measurement drives progress and accountability. As an organisation progresses in its D&I maturity, the metrics for measurement should similarly develop in order to help ensure that they are continually improving.

Key observations from the research project

- Continuous improvement is not just "a nice to have", but is essential in ensuring that the D&I agenda remains relevant to where the organisation is currently at and reflects what senior leaders and employees want.
- Established action plans need to be regularly reviewed against D&I goals.

(continued)

(continued)

- Review and measurement of progress against D&I goals allows the organisation to understand what it should start, stop or continue doing.
- Sharing and learning from leading practices, both within the sector and outside it, will help improve an organisation's D&I journey.

5.9 Message to the CEO

Committing to the principles associated with the RICS IEQM is a valuable first step in showing that you and your organisation are actively engaged in D&I. Formally committing to a standard may help to improve your results by enabling benchmarking and sharing and comparing leading practices with other organisations so that you can see areas for improvement.

In the UK, over 177 organisations have now signed up to the IEQM – many of these are international companies, therefore the IEQM values are already being "lived" in many corners of the globe. A full list of those organisations signed up to the IEQM is available on the RICS website.[3]

Adopting an approach based on the IEQM principles is a very effective way to start. Some may need to be adapted depending on the individual country and cultural differences, in which case some alternative criteria may have to be applied. While it is appreciated that not every company can adopt all aspects associated with D&I protected characteristics, simply developing a policy using the six principles will have a huge collective impact globally.

Every CEO should therefore take time to read through the key observations and findings of each of the principles and decide what works best for their own organisation, board and employees and seek their own professional advice so that the organisational workplace can be an inclusive environment where everyone can bring themselves to work every day.

Committing to a standard that is independently assessed, such as NES or the IEQM, is an invaluable way to demonstrate to your organisation at all levels – but especially to employees – that you take the D&I agenda seriously.

Notes

1 RICS & EY (2017). *Building Inclusivity: Laying the Foundations for the Future. RICS Inclusive Employer Quality Mark.* Available from: https://ccsbestpractice.org.uk/wp-content/uploads/2017/08/RICS-Building-Inclusivity-2016.pdf [Accessed 30 September 2018].
2 EY (2018). National Equality Standard. Available from: www.nationalequalitystandard.com/ [Accessed 30 September 2018].
3 RICS (2018). Available from: www.rics.org [Accessed 30 September 2018].

6 How to successfully implement a D&I strategy

6.1 Introduction

The starting point is, of course, recognising that the organisation can do more to embrace D&I. Depending on the size and scale of the organisation, plus the country or countries the organisation covers, this may be more or less complex to implement. What is also important on any journey is to understand where you are now and where you want to get to, and when, in order to set the parameters for that journey and the criteria against which you will be measured.

Each CEO or leader in the organisation needs to set the organisational aspirations and cultural environment for employees. Defining a strategy can be a challenge, but implementing a strategy is much more difficult, and very much depends on the pre-conditions for success.

6.2 The case for change

Being clear about the case for change is really important, and this has been set out throughout the preceding chapters, but using the National Equality Standard Business Case Drivers as a basis, these can be summarised as follows.

"I have to be honest with myself—I could do more.
I could be a bigger blot of ink in order to stain more stuff."

Financial performance

Improved financial performance has been linked to embracing D&I within an organisation. Organisations with greater levels of D&I outperform those with less, with average profitability up 27% and customer satisfaction up 39%.

There is increasing evidence that companies with women on their boards outperform those without by 25%. Companies with one or more women on their board[1] deliver higher than average returns on all key measures, such as 42% greater return on sales, 53% better return on equity and 66% higher return on invested capital.

Also, those organisations with increased racial diversity at board level bring in 15 times more sales revenue on average than those with lowest levels of racial diversity. Board diversity brings returns on investment that are 53% higher, on average, for those in the top quartile against those in the bottom quartile, with profit margins being 14% higher in those companies that are the most diverse. Companies in the top quartile for racial and ethnic diversity are 35% more likely to have financial returns above their respective national industry medians.[2]

Legal considerations

Workplace discrimination and harassment cases have been increasing in terms of numbers and payout amounts in recent years. The amount of legal and management time and reputational costs also need to be taken into account, as well as employee perception and commitment levels when such cases are prevalent. Increasing the focus on D&I across the organisation and improving awareness of all employees, but especially line managers, will improve both costs and morale.

Governments are continuing to expand laws and regulations around D&I, particularly relating to the fair and equal treatment of women and other disadvantaged groups in aspects such as pay and working environment. Gender pay gap reporting is becoming a legislative requirement in many countries, and is widely adopted in over 144 countries (see Chapter 3, subsection: Gender pay reporting). Countries where legislation for reporting applies focus closely on the issue, and this will invariably affect people's decisions in selecting employers.

Market impact

While D&I in its fullest aspects cannot be implemented in every country around the globe, the increased globalisation of businesses is creating a growing need for an inclusive workplace environment. Analysis has shown that those organisations that embrace diversity in their leadership have significant market advantage, with 45% more likely to improve market share and 70% more likely to capture a new market.

Reflecting clients' organisational D&I, and being able to clearly demonstrate a D&I strategy in action, will unquestionably lead to better organisational alignment and more work because client organisations like and respect those that mirror their own values. Diverse boards deliver stronger corporate oversight,[3] resulting in less market fraud[4] and a better reputation within the marketplace. Customer satisfaction and earnings increase when workforces reflect the broader population, while loyalty increases by making a public commitment to D&I.

"I appreciate your diverse collection of trinkets. I too have diverse tchotchkes."

The people dimension

Employees who think their organisation is committed to D&I are more likely to stay: they feel engaged,[5] experience less absenteeism, and are 80% more likely to perceive they work in a high-performing organisation.[6] Research involving over 50,000 employees shows that engagement is closely linked with loyalty to their managers, and engagement is highest where their managers show commitment to strong diversity.[7]

Promoting D&I will increase employee motivation, and staff retention as a result, with statistics evidencing 20–50% of employees intending to stay with their firm as a result of the D&I environment. In a survey of 200 European Companies, 62% said that diversity programmes had helped them to attract and retain highly talented people – an important consideration when the war for talent is exacerbated by falling birth rates in Europe, leading to a projected 24 million shortfall in workers aged 15–65 by 2040. Attracting and retaining more women in equal numbers to men would help reduce that gap to 3 million.

Finally, 72% of companies that have diverse teams report increased earnings by, on average, 10%. Highly engaged organisations have been shown to have an increase of 28% in earnings per share and an improvement in operating income of 19%.[8] In addition, they are generally more innovative and better at problem-solving. Diverse teams are more effective at solving complex problems than highly qualified expert teams.[9] Differences in perspective and experience enable diverse teams to generate more varied and plentiful idea combinations, leading to higher creativity and the out-of-the-box thinking that is crucial to innovation.[10]

Diverse work teams produce results that are six times higher than teams that are less diverse.[11] Organisations with greater levels of D&I outperform those with lower levels by 22% in increased productivity and 22% lower staff turnover,[12] and inclusive teams make better business decisions up to 87% of the time.[13]

Therefore, embracing D&I simply makes business sense, and makes for a healthy workplace culture as well as a winning organisation.

6.3 Business areas to consider

It is important to think about where change will be needed across the organisation and to what degree, depending on the strategy proposed and what the organisation currently has in place.

Self-assessment against the RICS Inclusive Employer Quality Mark (IEQM) or an independent assessment using the National Equality Standard (NES) are two acknowledged ways to understand where you are in terms of the organisational baseline and what needs to change to get to leading practice.

The areas of the business to consider aligning to the six IEQM principles (detailed more fully in Chapter 5) are as follows:

1 **leadership and vision** – strategy, setting priorities, culture, policies and practices, governance and management responsibilities;
2 **recruitment** – advertising, job descriptions, attraction, recruitment, staff on-boarding and induction;
3 **staff development** – performance monitoring, career progression, learning and development, targeted training, bias training and awareness;
4 **staff retention** – staff networks, mental health, wellbeing and wellness, flexible working, accessibility, workplace, caring responsibilities, coaching arrangements and mentoring;
5 **staff engagement** – communications, staff engagement surveys and approaches;
6 **continuous improvement** – quality and assurance assessments, benchmarking and regular reviews.

All of these areas need to be considered in detail in implementing D&I across the organisation.

6.4 Outlining the three key elements for implementation

The three key elements for successful implementation outlined below are adapted and summarised from the book *Making the Difference*, originally published in Dutch by Gretha van Geffen,[14] and provide a framework for us to explore a structure for implementing D&I in an organisation:

1 setting the vision and strategy;
2 the people dimension;
3 supporting process and systems.

We will now discuss each element in more detail.

Setting the vision and strategy

When it comes to a corporate vision of D&I, two currents can be identified. The first current is based on the rational view that "the business case needs to make sense", and the second current is based on righteousness. The latter is the oldest view, and originated in the United States. The United Kingdom is also strongly influenced by this view, and to a lesser extent, so are other countries.

The difference between the two currents is clear. The current that sees D&I as a business case puts business interests first and assesses the business opportunities that follow from actively pursuing a D&I strategy – in other words, how can a D&I strategy contribute successfully to the corporate goals and objectives of the organisation?

The current that feels that D&I should be based on righteousness focuses on social inequality and discrimination. Organisations taking this view usually also reflect it in their core values and business principles.

Both views make sense, but a hybrid seems ideal. Having a D&I strategy simply because it drives business performance does not feel right, and is likely to hurt your reputation among your stakeholders in the long term. Successful organisations these days are value-driven, and they need to be honest and authentic. Therefore, reference to D&I should also be embedded in the organisation's core values and business principles.

For vision development, it is important to take the following starting points into account:

- Vision development is not about telling people they are wrong or right.
- Being aware is not enough to change behaviour.
- It is not sufficient that the organisation "obeys the law".
- Leadership on D&I has to start at the very top of the organisation in setting the vision.
- You must link D&I to the business, particularly to innovation and human resource management.
- You need to define in your corporate objectives what D&I will bring to the table.
- "Ban mechanisms" should not be seen as a cause for behaviour, but as a challenge to change behaviour;
- You should think through the business consequences of the corporate vision of D&I.

Setting the strategy for D&I

The strategy that is set based on the agreed vision for D&I provides the overall scope and direction of the organisation through specific actions the organisation must take to address the business goals for D&I while achieving a competitive advantage.

The strategy looks at how the different elements of the business come together around D&I and how they impact one another in terms of the people dimension and supporting processes and systems.

Organisational culture

It goes without saying that changing and influencing the culture of the organisation are among the hardest things to do. The reason for this is that a culture is being formed by an amalgamation of values that employees in an organisation commonly share, and how the leadership of the organisation behaves is most important. It has to start at the top, with the organisational leaders walking the talk. This means that the culture is part of the organisation's DNA and is being reflected in how employees are being hired, managed and developed, and how people leave the organisation.

Since this is a practical book with ready-to-use solutions, we will introduce a five-step approach for changing organisational culture. This approach has been developed by Wharton University of Pennsylvania with its Nano Tools for Leaders® series,[15] and has been successfully implemented in many different companies throughout the years.

STEP 1: ASSESS THE CULTURAL VALUES

The first step in culture change is knowing where the current culture stands – that is, what employees believe the organisation's current values are. This will allow you to get a good idea of how much change is needed, how you can enable accountability, and what abilities exist to track culture change more precisely over time.

STEP 2: INTENTIONALLY ALIGN CULTURE, STRATEGY AND STRUCTURE

Be sure that the culture change fits with the organisation's business strategy and that both fit with your organisation's structure (its formal systems and policies). Reconsider formal reporting relationships, job descriptions, selection and recruiting practices, performance appraisal, reward or compensation structures, and training and development.

STEP 3: ENSURE STAFF AND STAKEHOLDER PARTICIPATION

Change cannot succeed without the meaningful involvement of many people throughout the organisation. Participation can range from individually offering ideas, solutions and reactions to concepts to taking part in team meetings to design and build the new culture and organisational structure. Use a balanced approach, keeping in mind that input from a wide range of people can generate excitement and motivation to change. However, make sure that a separate change structure is in place (for example, a change sponsor or change committee) that can make timely and clear decisions to prevent an ambiguous vision or delays to key actions.

STEP 4: COMMUNICATE AND DEMONSTRATE THE CHANGE

Communicate and demonstrate the change, again and again and again and then . . . again. Use words and actions to convey the vision of the desired future, and repeat the message more often than feels necessary. Role-model appropriate behaviours by demonstrating constant commitment to the future state and provide a vivid image of what it will look and feel like. Use new language or metaphors to create memorable images. Clear and visible dedication to change has been found to be one of the key success factors in culture change.

STEP 5: MANAGE THE EMOTIONAL RESPONSE

Leadership effectiveness in times of change is critically related to the use of emotional intelligence. Employees' emotions have a strong influence on how they approach change, and you need to be as analytical and strategise as much about their emotional messages as their cognitive ones. Pay attention to and read others' emotions, and empathise and engage in perspective-taking to better predict how employees will respond to change. Manage the anxiety, periods of anger and need for emotional regulation that can naturally arise at critical points in a culture change.

The above is a generic approach for change management. Therefore, using this approach for driving D&I means that special attention needs to be paid to, for example, "ban mechanisms", majority versus minority, and the diversity-friendliness of formal and informal rules and norms. Also, involve all sorts of people from all parts of the organisation to help in defining the culture, but also to act as ambassadors for change.

Engaging key stakeholders and ambassadors for change

Engaging employees and creating a culture of engagement in the organisational journey is essential, with the mantra of: listen, talk, act. Also, recognise that it is important to be prepared to tackle some of the uncomfortable truths that may emerge. This is why engaging key stakeholders and ambassadors for change is an important starting point in embedding D&I.

Being able to show the added value of a D&I strategy to stakeholders (the leadership and management team, staff, customers and suppliers) quite often starts with making stakeholders aware of the issue. This can be done in many ways, such as:

- actively engaging with the corporate network, specifically on the topic of D&I;
- sponsoring events where D&I is positioned as a theme;
- organise D&I-related events for stakeholders;
- try to win prizes or sponsor awards;
- encourage people in the organisation to speak about D&I at events;
- publish about the topic in relevant newspapers and magazines.

As soon as awareness is raised, it is possible to start showing the added value of D&I. Not every stakeholder will be equally interested in or sensitive to a topic such as D&I. Therefore, be careful in choosing the stakeholders to engage on D&I.

After this group of stakeholders has been identified, the next move is to investigate how a D&I strategy will address your stakeholders' needs and how this will add value for them. In this case, it is very important to reflect and honestly assess whether value truly can be added and whether your organisation can deliver on the issue of D&I.

*"It's time to honestly reflect
and assess true value."*

*"True, but also do something a little bad
like send a passive-aggressive email."*

Commitment from the top

The organisational culture has been discussed above, as well as the role leadership plays in defining and living the culture. Since implementing a D&I strategy can be sensitive, especially when an organisation is international and operates daily in various countries, it is critical for success that those at the top of the organisation openly commit to the strategy, and preferably one person from the top team should be made the sponsor for implementation.

To boost implementation, it is furthermore relevant to identify employees across the organisation who will be supportive and help you create commitment. Employees may be supportive because it is in alignment with their (political) beliefs or background. It is also possible that some employees may have experienced themselves or people close to them having been banned or excluded, or that some of your employees have been working in diverse organisations prior to joining yours.

Moreover, it is important to realise that quite often the level of resistance within the management of your organisation may exceed that of employees. Unlike, for example, the introduction of quality management or a reorganisation, this is especially significant in the implementation of a D&I strategy. The rationale behind this is that employees may consider that D&I is an issue for the rest of the staff, and not necessarily for them. Research has identified six major sources of resistance that you should be aware of:

- Fear for change makes managers close ranks to protect their own positions.
- Managers and employees feel that they have not done things right previously.
- Managers state that for the time being they cannot, or do not need to, hire new staff, as if D&I is only about bringing in new and different people.
- Managers consider it as something truly additional, over and beyond everything else. Therefore, for them, it is low on the list of priorities.
- Managers state that they already have a good team, and change will cause negative disruption.
- Managers quite often feel that doing nothing is less risky than doing something.

When it comes to the above, management teams must realise that D&I impacts everybody in the organisation, and maintaining too general a vision does not make the link with the daily business, and leads to failure. It is therefore helpful to acknowledge that resistance exists, and extensive communication about why D&I is important is critical to overcome it. Work with the willing, share responsibilities and offer enough support to your people.

Summary

- **Establish the vision** – As the leader, you will have a clear vision on D&I and how this vision is linked to the corporate vision and mission. What is important now is to set out the aspirational vision for how D&I will be embraced by the organisation in a way that reflects the values and culture of the organisation, as well as the practical limitations.
- **Establish the organisation's baseline** – It is important to establish some key facts around what the organisation is currently doing. You need to achieve an honest understanding of the organisation as it currently is in terms of D&I – the best way to do this is independently from your employees and customers, as well as from the business metrics.
- **Set the strategy** – Once you have set aspirations for the organisation, it is possible to set the strategy as your organisation moves towards a culture characterised by change-mindedness, an openness to diverse behaviour and different ways of thinking. It is important that you can show the added value of a D&I strategy to your stakeholders, primarily your clients and business relations, as well as your managers and employees. Make sure you allocate resources (and budget) to support delivering against the agenda you set.
- **Set the organisational cultural tone and show real commitment** – You and your colleagues at the top of the organisation need to truly commit to the cause and embrace the strategy wholeheartedly. This is

about creating the culture and environment you wish for the organisation. Be clear about what D&I means for the organisation and how it will be incorporated. Avoid leaving aspects open to interpretation (or misinterpretation!). Flaws or inconsistency in messaging and/or behaviours will be quickly seen by employees as a lack of commitment and the organisation paying lip service to the strategy.

- **Engage key stakeholders and change ambassadors** – Work to identify key stakeholders and change ambassadors within as well as outside the organisation that can give you honest counsel and support, as well as widening the capacity to deliver.
- **Review governance and management** – Putting in place new, or adjusting existing, governance and management responsibilities is important in order to provide the right structural environment for the D&I Strategy to be successful.
- **Measure progress** – Setting priorities for the strategy with clear responsibilities and accountabilities along with timelines will ensure that the business monitors progress during the transitional change. Remember to take time to celebrate and publicise success.

The people dimension

Employees

Implementing a D&I strategy is not an easy thing to do and much depends on the change-mindedness of your leaders, managers and employees, and how open they are to change related to D&I. However, this is manageable, based on clearly defined D&I skills and competencies. Prof. Dr emeritus Wasif Shadid,[16] a well-known researcher in this field, therefore divided the diversity competency in three different categories: motivation, knowledge and skills. Each category is further detailed below.

MOTIVATION

The motivation of employees is formed due to a combination of the following factors:

- Employees need to be open to actively engaging with other people/colleagues and maintain relationships with them.
- Employees need to be self-confident: they may be afraid of losing their own identity, making a bad impression or being dominated.
- They need to know why a change is needed and they are going to be rewarded.
- They may have pre-assumptions.
- They may have had bad experiences in the past.
- Sometimes employees want to maintain a certain distance towards other groups of people. This can tie in with religion, class, culture etc.

KNOWLEDGE

To a certain extent, everybody needs to be knowledgeable about D&I. It is important to raise appropriate expectations and avoid misinterpretations. Knowledge of communication and (non-)verbal behaviour in different (national) cultures is key for success, as well as proper knowledge of norms and values.

SKILLS

Skills are primarily about how you communicate and behave. An individual's D&I can be scored via the matrix in Table 6.1. Originally this matrix was intended to score international employees, but it can also be used to assess D&I competency.

Table 6.1 D&I competency scoring matrix

Category	Indicator
Intercultural sensitivity	**High scores indicate the ability to:**
	Recognise one's own cultural values and those of others
	Discover and appreciate other people's values
	Recognise multiple perspectives on an event or behaviour
	Pick up and recognise verbal and non-verbal signals in communication and needs of listeners
Managing uncertainty	**High scores indicate the ability to:**
	Feel comfortable with new and unpredictable situations
	Respond flexibly and work effectively in new cultural situations and with new people
	Not be stressed or nervous when faced with different beliefs, habits and ways of communicating
Intercultural communication	**High scores indicate the ability to:**
	Adjust one's communication style to the communication needs of people from other cultures
	Adapt ways to effectively explain, describe and convey messages to people from other cultures
	Modulate different degrees of directness in difficult communicative situations
	Take care not to cause loss of face when communicating about negative events
Building commitment	**High scores indicate the ability to:**
	Establish relationships with a diverse set of people and get them committed to a shared task
	Understand organisational politics
	Encourage exchange between people and stimulate them to contribute
	Take the lead while at the same time keeping others on board

Leaders and managers

Leaders and managers are very relevant to successful implementation of D&I across the organisation, in particular when it comes to dealing with D&I issues within their own teams or departments. There are five ways to respond to these kinds of issues:

- **Dominance** – The manager decides how the team works and communicates together.
- **Adaptation** – We must successfully work and communicate together. So why don't we do it your way?
- **Avoidance** – Driven by, for example, fear of conflict or political correctness, the manager does not address issues properly.
- **Compromise** – We try to work and communicate together in a collaborative way. Today we work according to my way, and tomorrow your way of working prevails.
- **Synergy** – The manager actively stimulates joint development of a new way of working and communicating together.

Which of these responses makes most sense will depend on the situation, but it is important as a manager, not to make the following mistakes:

- not asking questions or listening to your team, suppressing differences or letting the strongest prevail;
- failing to encourage everyone to be a leader on D&I, as often the most influential people and the best ideas come from the lowest levels in the organisational hierarchy;
- being (perhaps unconsciously) afraid of being accused of discrimination;
- not offering clear guidance when it comes to conversations about D&I.

Instead, it makes sense to take the advantages of D&I as a starting point for action, to see a D&I-related incident as an opportunity for change and communication, to accept living with a certain level of uncertainty, and accepting that D&I is always a quest: you do not have to be perfect as long as you keep your goals and objectives in sight.

Competencies and skills

Competencies and skills have been touched upon already, but they deserve more explanation because in today's world, it is less about job descriptions and more about competencies and skills, due to the changing nature of work.

Competencies and skills can be assessed at personal level, team level, and organisation level, and at all those levels competencies and skills need to be defined that reflect not only the D&I strategy, but also how these competencies and skills contribute to the corporate goals and objectives. Apart from the three levels referred to above, there are also groups in society with specific competencies and skills – for example, in language. Depending on the work to be done, people can be hired because they belong to that specific group in society.

When it comes to integrating D&I competencies and skills in your organisational competency framework, it is important to avoid the following mistakes:

- assuming that if someone belongs to a specific group, they will have specific competencies and skills;
- creating human "copies" instead of focusing on individuals' strong points, especially in relation to team competencies;
- thinking that defining competencies and skills is a "neutral" activity.

Training

Giving employees (including leaders and managers) the right tools to understand and implement the D&I strategy is essential. An effective way to do this is through training on an individual, team or peer basis.

"That tool's for looking at stuff, and this magical one's for implementing against the D&I strategy."

Providing knowledge about the tools available through effective training provides an immediate spur towards D&I goals, particularly in terms of providing a better understanding of ethics or unconscious bias, for example.

Training does not have to be expensive to implement or undertake, but it does need to be effective, and should be evaluated and measured in terms of its effectiveness and adapted as appropriate to suit the organisation's needs.

Summary

- **Communicate the strategy** – Everyone in the organisation needs to understand the vision and strategy that are being set with regard to D&I, and most importantly, what it means for them. Think about key motivators, sharing knowledge and assessing D&I skills.
- **Give your teams the tools** – It is important that your teams have the knowledge and skills to deal adequately with D&I matters. Providing support processes and systems, but particularly a specific contact in HR, will help the implementation.
- **Leaders and managers must walk the talk** – Your leaders and managers need to recognise and address D&I dynamics and act in a way that shows the advantages of D&I on a daily basis. People look to each other, and particularly upwards to their line manager and leaders, to see the behaviours for success that they need to emulate. Give permission to each other, especially as leaders, to pick each other up when behaviours or actions are observed that do not reflect what has been agreed.
- **Embed the D&I competencies** – It is important that your organisation has a clear insight into the D&I competencies and skills of staff in combination with other relevant business competencies and skills. Make sure that D&I competencies are embedded and embraced throughout the organisation and across the employee lifecycle from staff recruitment to development, retention and engagement.
- **Provide adequate training** – Training will help disseminate knowledge and gain momentum. Ensure there is a sufficient budget for training, as this will help expedite the pace of change and implementation of the D&I strategy.

Supporting processes and systems

Capturing the D&I principle in strategic human resource management, communication, marketing and management style helps to engage those who do not really buy in to the idea of D&I. One advantage of capturing D&I in corporate instruments is that D&I becomes policy and is less arbitrary.

It is not possible to discuss every system or instrument in the organisation, so we will explore some examples. The most obvious ones are instruments for human resource management. How diverse and inclusive are your corporate instruments? These may include:

- job advertisements;
- tests and assessments;
- recruitment agencies selected and used;
- forms;

- flexible working hours;
- competence management;
- job descriptions;
- personal development plans;
- introduction days.

All these instruments serve the overarching purpose of discovering and utilising talent in your organisation.

Marketing and communication instruments relate to, for example, a focus on target groups, communication channels, brochures, promotion and products. Not every organisation consciously thinks about the management style that is being used, but in general, an inspiring, coaching and facilitating management style will help with the implementation of a D&I strategy.

Issues to think of include:

- how instruments can positively contribute to your D&I strategy;
- out-of-the-box thinking is necessary to capture D&I in your instruments;
- instruments are key for your image and reputation.

Workforce

D&I should not just be reflected in the lower levels of the organisation, but needs to be visible from top to bottom. To achieve this, there are a number of things you can do, such as:

- adopting positive discrimination – or, in a weaker form, preference policies – to purposely increase numbers from specific groups in your organisation;
- changing your recruitment channels and networks in which you operate;
- installing and training diverse interview panels;
- recruiting people based on values and team composition;
- setting targets for your managers;
- working with instruments like mentoring and job shadowing.

A strong business case is needed if you want to change the make-up of your organisation – a business case that reflects a strengthening of your organisation's insights and relationships with your diverse client base and environment.

Management assessment

The business case can set the baseline, but what is more important is the need to focus on actions and measures. Actions and behaviour by the (top) management of your organisation will be copied by the rest of your workforce. Therefore, it is crucial to think about actions and behaviour and how desired behaviour can be captured and institutionalised. Ways to do this relate to:

- setting measurable objectives for client and team satisfaction, growth in identified target groups, innovation etc.;

- defining competencies and skills;
- including D&I in your mid-year and year-end reviews;
- examining how new management staff are being recruited and developed.

It is not enough for managers to have their hearts in the right place or to reward only managers that perform in the same way you do. Also, be aware that everything is interlinked, which means that, for example, recruitment, reviews and development need to be aligned.

Summary

- **Define processes** – It is important to look at the existing organisational processes that are likely to be impacted by the D&I strategy. Many will be in HR, but some will be in the governance, management, finance, marketing and communications areas. It is important to look across the whole organisation and think of the people touch points (both employees and customer) in the business, outlined as part of the six IEQM principles (detailed in section 6.3: Business areas to consider), and think about the processes in the business that need to change.
- **Enabling systems** – With so much of business administration being digitally enabled through technology and data, it is worth considering the systems and tools that can be put in place to help enable this agenda, including recruitment, guidance notes, staff communications, access to policies and procedures, complaints handling, appraisal monitoring, data capture leading to dashboarding progress etc. Think about how systems can be best utilised to help enable the vision and strategy.
- **Supporting departments** – D&I principles need to be embedded in instruments related to strategic human resource management and marketing and communications. It is important that key support departments such as these are ready to support the roll-out and implementation of D&I across the organisation. This should be an exciting time for the organisation, and getting the messaging clear, backed up by the right tools, processes and systems, including information technology, will make for better implementation.
- **Changing behaviours** – It is important that D&I is reflected at every level of the organisation, from the top to the bottom. Showing that (top) management are being assessed and rewarded based on actions and behaviour related to D&I really shows that D&I is being taken seriously and is part of the organisational DNA.

6.5 Message to the CEO

There is a lot of evidence for the rationale for organisational change towards D&I in terms of business case statistics, and there is strong supporting evidence, some of which has been outlined in this chapter. However, for many CEOs there is a strong recognition that addressing D&I is *the* right thing to do. Full stop. In tackling the war for talent and getting the organisation fit for the modern working world, D&I is key. Lack of diversity is the real issue for organisations, and the solution includes opening up the barriers to greater inclusivity. Setting the right vision and strategy are the essential starting points, as is having a thorough understanding of where the organisation currently is and where it wants to get. Using assessment approaches such as IEQM or NES is an effective way to understand your organisation's performance compared to peer organisations, and also those in other sectors in the case of NES.

The key is to bring the whole organisation with you on the journey, from top to bottom, and to ensure that everyone understands the rationale and change required through the business case – essentially, why D&I matters and that non-conformance is not an option.

"Everyone come along—from the tiniest theodolite to the giantest real estate agent!"

Implementation is about people, processes and systems, and will affect and embrace many areas of the business. Do not underestimate the importance of the people dimension on the journey.

When you want to implement a D&I strategy, adopt a structured approach as detailed in this chapter, assume accountability and delegate responsibility to key individuals from across the business. Also, make sure that your D&I strategy is aligned with your corporate strategy and that someone from your board is made responsible for success, as this will show real commitment from the top.

Of paramount importance is starting the D&I journey, focusing on actions and measures, and embracing change. Only 51% of firms in the sector regularly review progress towards D&I goals, and 79% of firms with a D&I action plan in place reported regular reviews of progress towards D&I goals.[17] Wherever you are on your own organisation's D&I journey, there is always room to improve and to evolve to tackle the war for talent. As the case for change has outlined, addressing D&I simply makes good business sense – but it is now time for action!

Notes

1 Carter, N. M., Joy, Lois, Wagner, H. M. & Narayan, S. (2007). *The Bottom Line: Corporate Performance and Women's Representation on Boards.* Available from: www.catalyst.org/knowledge/bottom-line-corporate-performance-and-womens-representation-boards [Accessed 30 September 2018].

2 Hunt, V. Prince, S., Dixon-Fyle, S. & Yee, L. (2018). *Delivering through Diversity.* New York: McKinsey & Company. Available from: www.mckinsey.com/~/media/mckinsey/business%20functions/organization/our%20insights/delivering%20through%20diversity/delivering-through-diversity_full-report.ashx [Accessed 30 September 2018].

3 Adams, B. R. & Ferreira, D. (2009). Women in the boardroom and their impact on governance and performance. *Journal of Financial Economics* 94(2): 291–309.

4 Cumming, D., Leung, T. Y. & Rui, O. M. (2012) Gender diversity and securities fraud. *SSRN Electronic Journal* 58(5). Available from: www.researchgate.net/publication/256035829_Gender_Diversity_and_Securities_Fraud [Accessed 30 September 2018].

5 Corporate Executive Board (2004). *Driving Performance and Retention through Employee Engagement.* Available from: www.stcloudstate.edu/humanresources/_files/documents/supv-brown-bag/employee-engagement.pdf [Accessed 30 September 2018].

6 Deloitte (2012). *Deloitte 2012 Global Report.* Available from: https://public.deloitte.com/media/0564/pdfs/DTTL_2012GlobalReport.pdf [Accessed 30 September 2018].

7 Corporate Executive Board (2004). *Driving Performance and Retention through Employee Engagement.* Available from: www.stcloudstate.edu/humanresources/_files/documents/supv-brown-bag/employee-engagement.pdf [Accessed 30 September 2018].

8 Towers Perrin (2008). *Closing the Engagement Gap: A Road Map for Driving Superior Business Performance. Towers Perrin Global Workforce Study 2007–2008.* Available from: https://c.ymcdn.com/sites/www.simnet.org/resource/group/066D79D1-E2A8-4AB5-B621-60E58640FF7B/leadership_workshop_2010/towers_perrin_global_workfor.pdf [Accessed 30 September 2018].

9 Hong, L & Page, E. S. (2004) Groups of diverse problem solvers can outperform groups of high-ability problem solvers. *Proceedings of the National Academy of Sciences* 101(46): 16385–16389. Available from: www.pnas.org/content/101/46/16385 [Accessed 30 September 2018].

10 Higgs, M., Plewnia, U. & Ploch, J. (2005) Influence of team composition and task complexity on team performance. *Team Performance Management* 11(7): 227.

11 Page, E. S. (2007). *The Difference: How the Power of Diversity Creates Better Groups, Firms, Schools, and Societies* (The William G. Bowen Series). Princeton, NJ: Princeton University Press.

12 Cumulative Gallup Workplace Studies.

13 Larsen, E. (21 September 2017), New research: diversity + inclusion = better decision making at work. *Forbes.* Available from: www.forbes.com/sites/eriklarson/2017/09/21/new-research-diversity-inclusion-better-decision-making-at-work/#65b989004cbf [Accessed 30 September 2018].

14 Van Geffen, G. (2010). *Making the Difference: The Critical Success Factors for Diversity Management.* Champaign, IL: Common Ground Publishing.

15 Wharton University of Pennsylvania (2016). Wharton's new app: Nano Tools for Leaders®. Available from: https://executiveeducation.wharton.upenn.edu/thought-leadership/wharton-at-work/2016/04/whartons-new-app/ [Accessed 30 September 2018].

16 Diversity Research Focus, W. Shadid (2012). Interculturele communicatie. Available from: www.interculturelecommunicatie.com/engels.html [Accessed 30 September 2018].

17 RICS & EY (2017). *Building Inclusivity: Laying the Foundations for the Future.* Available from: https://ccsbestpractice.org.uk/wp-content/uploads/2017/08/RICS-Building-Inclusivity-2016.pdf [Accessed 30 September 2018].

7 Perspectives

7.1 Introduction

This book is about people.

People who have a job to do and who want to be the best at doing that job. People anywhere in the world. People who may be working in the real estate sector, or who may not be. People who want to do their best at work doing a job they love. People who may not be able to do that job to the best of their ability because they are trapped by the dictates of society, constrained by social norms, or simply part of a company which neglects a diverse and inclusive culture. People who want to change things, but who cannot. Or equally, people who have the power and capacity to change things, but who may not know how.

The Sequoia National Park in California is home to the most awesome redwoods that have throughout the ages survived world wars, drought and horrific fires. Yet they stand proud and tall, living evidence that in the right conditions and environment it is possible to thrive despite adversity and against many odds. A tree branches towards the sky, seeking light, sun and air. Below the surface there is a myriad interplay of soil conditions, flora and fauna, drainage, nurturing those roots which will allow the branches and leaves to reach out higher to the sky. Below the surface is a living world of communication and interaction. That tree needs to be understood. It needs to be nourished and treated with respect. Given the right conditions – soil, light, water – it can, and will, flourish and thrive.

And that is how it is with D&I, and how it can help make people grow strong and tall.

We are all on a journey, and we all have a story to tell. That story influences our perspectives, both consciously and subconsciously. By writing this book, we as authors wanted to share our thoughts and experiences with you, to inspire you to think more about D&I. If we had left it there, it could have turned out rather one-sided. So we invited eminent leaders from around the world to also share their own unique perspectives with you.

We hope, that like us, you find the modesty of these stories awesome and the richness of experiences overwhelming. They all have a similar message:

Do not neglect what is beneath the surface, because it is that which makes people strong.

7.2 The time is now. D&I in this sector is woefully behind at a time when talent has never been more important

Arun Batra, CEO/founder, UK National Equality Standard

Arun Batra is the Chief Executive Officer and founder of the UK National Equality Standard, is an Associate Partner at EY, and has held numerous leading national diversity roles, including Diversity Director for the Home Office, Diversity Director for the Greater London Authority for two former mayors, as well as an advisor to the former Attorney General, Baroness Scotland.

He won the prestigious HR Consultant of the Year accolade for making "a significant difference to UK society", and has been repeatedly recognised as one of the UK's most influential Asians in the Asian Power List. In February 2018 Arun accepted an invitation from the Prime Minister to join a board to help tackle race disparity in the UK.

Diversity and inclusion has been something I have always been passionate about. It is in my DNA and is what I passionately believe in. It has helped define not only who I am, but what I now do.

I read law and am now the CEO or the National Equality Standard, or NES, which is proudly supported by EY LLP. The NES has proudly become a recognised standard across a wide range of sectors and organisations, and is something I established as a benchmark for organisations that truly wanted to demonstrate and improve upon the diversity and inclusion within their organisation.

Over the last four years I have had the privilege to get up close and intrinsically involved with the real estate and construction sector at a time where the sector was truly starting to embrace and tackle the diversity and inclusion within it. My lens has been unique, as my team and I have helped support RICS in setting up the Inclusive Employer Quality Mark (IEQM), in writing the report *Building Inclusivity: Laying the Foundations for the Future* (published in 2017) and in helping many organisations achieve and develop a path on the way to achieving the National Equality Standard.

Message to the CEO

So what message would I want to give a CEO or senior leader in this sector?

That the time is now. Diversity and inclusion in this sector is woefully behind many others, and yet the opportunities to attract, retain and grow talent have never been more important. With a well-publicised war for talent evident, people are invariably attracted to organisations that are seen to be positively creating an environment where everyone can be successful, can be themselves and can have a successful career.

"The time to grow talent is now!
Don't just water your talent with tea—this is a metaphor."

Many of the organisations from the sector are going through the NES as they believe it is not only the right thing to do as corporates, but for their employees and also to meet their clients' expectations about being the type of organisation that will be successful. Tackling diversity and inclusion simply makes business sense.

In the same way as achieving a standard for quality service delivery, holding client data or meeting legal requirements become an integral part of corporate business and demonstrate adherence in a world where trust, standards and regulation are key, so too is being able to demonstrate credibility through the achievement of a diversity and inclusion benchmark. The first rule in navigation is understanding where you are on the journey you wish to take. By understanding the levers you can adjust, whether a large global corporate or small SME, having an agenda to understand and improve diversity and inclusion within your organisation quite simply is a differentiator and makes business sense.

7.3 Create a workplace that is truly inclusive – so that everyone can be their authentic self

Antonia Belcher, Founding Partner, mhbc

Antonia Belcher has 40-plus years' experience in the construction and property industries and leads her own business that she formed in 2007. Transitioning in 2000/2003 in a very male-dominated working environment, where there was no history or visible LGBT influences to draw on, she presses for positive change for LGBT+ people in all business spheres, but especially in her chosen career path of surveying.

A Board Director of the Chartered Surveyors Training Trust, Diversity Role Models, a Liveryman of The Worshipful Company of Chartered Surveyors, and advisor to other organisations, she particularly seeks equal opportunities for women/LGBT+ people within the construction industry, which has of course been a sector of business generally slow to react to these developments. She was shortlisted in 2017 for the *Gay Times* "Success in Entrepreneurship" Gary Frisch award, and in 2018 was nominated in the NatWest British LGBT Awards – Inspiration Leader category.

As I write this, I have just been advised that I have been nominated among a group of "exceptional surveyors from RICS members around the world". The year 2018 marks the 150th anniversary of the Royal Institution of Chartered Surveyors, and to celebrate this milestone, RICS, of which I am a member, is showcasing the significant and positive impact that surveyors have made in society through those 150 years. My nomination is for championing LGBTQ issues in the surveying industry, and I am very proud to have been nominated.

As a trans woman who started her transition as the millennium began, I found myself having to convince my business partners at a very well respected West End consultancy that they should not ask me to leave because of the journey I was starting. Battling with the trauma I was causing within my family and social circles, the last thing I needed was to be out of work. Obviously, the omens were not good, especially with no LGBTQ influences evident at my place of work, an all-male equity partnership who had already made me sense that female involvement was not really wanted/expected at equity level, and probably not via the way I was intending. As it turned out, I was not told to leave, but it was more a decision out of commercial necessity rather than willing nurtured support, which has surprised me no end, given I have gone on to use my authenticity for so much good, both commercially and socially, and if you closely look at the changes now taking shape in the business world to be diverse in outlook and inclusive in approach. If only my old partners had realised that they were in effect ground-breaking, and could have made this decision with positive joy and encouragement rather than "'he/she' makes too much money to ask 'him/her' to go"!

I adore my profession, and the fact that I could plan to come to work as Antonia, and transition to achieve my full (and previously hidden) true female self has showed me that construction and property can be truly inclusive. I have been lucky in my transition, especially in not losing my family or work colleagues and being able to continue in life without too much interruption through that transition. Forming my own consultancy in 2007 with colleague surveyors from the firm where I transitioned further proved to me how inclusive surveyors can and want to be.

Message to the CEO

On this note, my strong advice to any CEO is to create a workplace that is inclusive, but make sure that the culture throughout all levels is truly inclusive, and it is not just a direction from on top. Ensure all management levels share this ideal and practise that pursuit of inclusiveness. Where you are steering a large enterprise, realise the power you hold to create change and instil a culture that searches out diversity, makes all forms welcome, and builds a business structure that will thrive on inclusivity. The larger you are, the more trans staff you will have, and if none are out, then your culture is not right yet, and needs urgent work. If you do not get it right soon, then those trans staff will leave you for a competitor who is making the effort and has realised it is a "no-brainer" to act now. Hidden trans workers in your organisation have so much more to offer in their "authentic" selves, and you are wasting that talent. Your act of allowing and nurturing them to be their true selves is a fantastic testament and metric of your business.

"Well, now I guess I'm truly INKlusive. Thank you."

Helping the SME supply chain that supports you, and helps you to look good, is a responsibility you should readily adopt, too. SMEs do not have the resources of the large corporates, and need help to find their way into the inclusive workplace.

7.4 Consider less conventional routes – implement real change, starting with education

Ciaran Bird, UK Managing Director, CBRE Ltd

Ciaran Bird became Head of UK Retail within CBRE in 2005, advising on a wide range of projects for many of his notable clients, including Marks & Spencer, Arcadia, New Look, Deloitte and Westfield.

Ciaran was appointed UK Managing Director in January 2013, and has responsibility for the management and growth of the UK business, while also continuing to advise his key domestic and international occupier and investor clients on their property strategy, both within the UK and Europe. Ciaran also sits on the EMEA Board and on the EMEA Executive Committee, and is responsible for Retail, Ireland, Client Care and Finishing First.

In Britain, we have an educational system with a strongly held view that university is the only route into the workplace. I'm Ciaran Bird, I left school at 16, and thinking I might be good at selling big houses while dreaming of becoming a professional rugby player, I fell into the world of commercial property. My first taste of the industry was in an office junior role at Fletcher King and later with Clive Lewis, before joining the renowned retail agency Dalgleish in 1990. Following CBRE's acquisition of Dalgleish in 2005, I became Head of UK Retail. Since 2013, I have been UK Managing Director of CBRE, a business with over 2,500 staff nationwide.

The fact I did not go to university nor had a conventional path into property has certainly not held me back. If anything, it has made me even more driven to work harder and seize opportunities as they arise. Right from the outset of my career I have been fortunate enough to work for some of the greatest UK retail brands, leading property companies and some truly inspirational people.

When rugby eventually turned professional in 1995, I was given the opportunity to play for London Irish while continuing to work. I was challenged with working in the office in the morning and training in the afternoon! This taught me very early on that the values and behaviours of any successful team are intrinsically the same, whether in sport or business. On the rugby field, teamwork meant respect, looking out for each other, and fostering a culture of hard work and discipline. This has stayed with me, and I am pleased to say that at CBRE we have instilled the same "teamship" values and have proven that if you trust and work collaboratively as one team, everyone benefits, especially the client.

So, recognising the power of teamship and that taking unconventional routes into a profession can have its benefits, I was determined to establish the CBRE Apprenticeship scheme in 2013, both as an investment in the next generation and as a way of championing diversity within the property sector. The first of its kind in the property sector, it was introduced to help widen the talent pool and provide an alternative route into the property industry. We now have over 50 apprentices across the business, and one of the proudest moments for me has been the successful graduation of our inaugural group of apprentices who have achieved Bachelor of Science with Honours degrees in Real Estate Management from Kingston University. It is worth noting that a third of the apprentices achieved first-class honours.

I particularly like seeing the impact these young individuals have on others in the company. Their work ethic, determination and spark are inspirational, and consequently, they have been integral to creating and driving change in the industry by sitting on panels and chairing industry groups. I believe they add a dimension to our workforce which will be the life and blood of the organisation for years to come and help shape CBRE's future.

On the subject of diversity more generally, we are still on a journey and there is so much more that needs to be done. Gender imbalance is an issue which prevails across all industries, and property is no exception. While we have come a long way, we still find progress in this area across the entire industry to be frustratingly slow. I believe that a fundamental part of this change lies in education. One of my biggest annoyances is that schools, students, their parents and careers advisors do not really understand the property profession. When people think about property, they think about estate agents and construction. There is very limited awareness in schools about the fascinating complexity of property and the built environment and all the other areas and paths it entails.

We cannot expect a diverse pool of young people, and especially women, to enter a profession they have never heard of. We need to not just talk about a step change in the workforce, as more education is needed as to what opportunities are available. As part of our extensive Schools Outreach programme, we insist at CBRE that apprentices and graduates go into schools to talk about the property profession, but what I'd like to see is for this to go one step further. Together with our competitors, our professional body RICS and government, we need to be owning the agenda and working collaboratively to educate schools so people of all backgrounds are aware of the opportunities within the industry.

On a grassroots level, a cultural shift takes time. When we started our apprentice scheme five years ago, we had 200 applications, and only five of those were from women, but last year the balance was fairly evenly split. The CBRE Women's Network started 13 years ago and is now a 1,000-plus community, but in common with other industries, there is still a lack of senior women at board level. Furthermore, we need to do more to create an environment and foster a culture which women want to be part of, before and after having children. In a brokerage environment, this is not always easy, flexible working can be challenging in the depths of deal completion, but clients need to understand and adapt with us.

This year, CBRE met the criteria for the National Equality Standard, addressing no less than 49 criteria to achieve the recognition which embeds our diversity and inclusion ambition right through our business and HR processes, with mechanisms for continuous review and measurement of progress. We are the first real estate advisor to have achieved this. We are also signatories to the Inclusive Employer Quality Mark, pledging to continually review our employment practices and to hold inclusivity at the heart of what we do.

Message to the CEO

My final word and my message to other CEOs is this: diversity and inclusion is not about ticking boxes, you have to believe it and implement real change, starting with education. Go for industry accreditation, but go beyond that and look to drive a paradigm shift in culture and behaviour. Do not underestimate the commercial imperative of having a diverse and collaborative workforce. From my own experience, I know that if you get this right, you will be a more successful business, where people will thrive and will want to stay. The barriers to achieving this are challenging and there are no quick fixes, but the rewards and the ability to create something special are off the scale.

"Don't just tick me, Bro ... Unless you believe in diversity and inclusion too?!"

7.5 Look beyond the obvious, so people are judged based on outputs, and not perceptions.

Justin Carty, Senior Director, CBRE Capital Advisors Ltd | Investment Advisory

Justin Carty is a Senior Director within CBRE's Investment Advisory Team. He specialises in structuring public/private partnerships and joint ventures bringing together commercial viability, finance/funding and development expertise to enable the delivery of complex residential, retail and mixed-use schemes.

He has advised on more than £1 billion of debt and equity investment into residential and mixed-use schemes throughout England, sourced funding for developers seeking to bring forward strategic sites, and advised on the delivery of innovative public/private projects. His projects have a combined value of more than £10 billion. In addition, Justin is a member of CBRE's Diversity Steering Group, set up and leads the CBRE Multi-Cultural Network and is Vice Chairman of DiverseCity Surveyors – a cross-industry network for black, Asian and minority ethnic (BAME) professionals

Dear CEO,

I'd like to ask you a couple of questions if I may.

What image comes into your mind when I describe the following persons?

Person A was privately educated, an award-winning student, and is now a senior director within the world's leading property advisory firm CBRE. This person's contribution to CBRE was recognised by being awarded the prestigious UK Business Advisory Award in 2017.

What do you think this person looks like?

Person B grew up on a council estate in south London. Raised in a single-parent immigrant family, he has been stopped and searched by the police on 19 occasions and experiences people holding on to their belongings when they walk by.

What do you think this person looks like?

Would it surprise you if I said that I am both Person A and B?

I think our experiences in life and in the workplace are governed by unconscious bias. I will elaborate further on this, but first let me introduce myself properly.

My name is Justin Carty, and I am a Senior Director within CBRE Investment Advisory. I have advised on a wide range of development, investment and funding mandates with a combined value in excess of £10 billion. My projects involve the delivery of thousands of homes, community and educational facilities and jobs. I also help my clients achieve financial returns.

I grew up in a single-parent family on a south London council estate and am of Afro-Caribbean descent. My mum single-handedly raised me and my sister. Growing up, we didn't have much money at all. My mother is a true hero, and taught us the values of education, hard work, respect and being proud of who you are and where you come from. She had the foresight to send me to private school, gave me intensive tutoring in the lead-up to the entrance exam, and used her persuasive skills to secure a bursary as there was no way we could afford the fees otherwise. She also taught me that I'd have to work ten times harder to be valued the same. How right she was.

Over the last three or four years I have been on a journey of self-discovery. I have always been acutely aware that as a black man I am, rightly or wrongly, perceived in a particular way outside of the office, but didn't realise this could also be the case within the office. At times, during my career I have felt that I may have been being paid less than my peers (while on paper, at least, being one of the strongest performers) and I have always considered that I have had to outperform my peers in order to be promoted. In my past I have also successfully challenged performance-related pay when I felt that the terms offered didn't reflect the outputs achieved. I value being at CBRE because I feel valued, but all of this has left me wondering, to quote the famous poet Ali G, "Is it cos I is black?"

On a day-to-day basis I never felt that I was treated differently, so I was left scratching my head as to why, when it came to pay and promotion, there seemed to be a bit of a struggle.

In 2015 I attended a course about unconscious bias. This is when the penny dropped for me. I realised that we all have unconscious bias. We all have perceptions about the type of people who do certain jobs, and this is shaped by life experiences, and in particular the media. I realised that I have unconscious bias as I assumed that nurses are always women and chairpersons are always middle-class, middle-aged white men.

I felt that unconscious bias may therefore have played a role in the way that I've been treated, as one wouldn't naturally expect a black man from a council estate to have a senior role in a large corporate company.

In 2017 I was promoted to Senior Director, and this was a very proud moment for me. I am the first black person to get to this level, which is a wonderful personal achievement. However, I am also slightly surprised that in today's society this is the case, and I often ask myself why don't more people from a BAME background come into this great profession?

Of course, I have had to work hard to get here. I was client care manager for two key clients, both of which I secured myself. I generated significant fees of over £1 million per annum, I mentored three junior employees, used to run the football team, sit on our diversity steering group, set up the multi-cultural network and won the UK Business Advisory Award. So, when I got to the promotion interview, I received feedback from HR to say that it was one of the highest-scoring promotion interviews over the last five years. So I guess you could say I absolutely got here on merit, and through lots of hard work!

Message to the CEO

My message to the CEO is that we need to look beyond the obvious and move towards ensuring that people are judged based on outputs, and not perceptions. My mission is to ensure that we attract, retain and enable the attainment of diverse talent, but without them having to outperform. It is important that the hurdles for promotion are the same for everyone.

Look beyond the obvious.

CBRE has made significant inroads towards enabling this, as evidenced by being awarded the National Equality Standard. It is proven that more diverse workplaces lead to happier employees and better financial outputs. The argument for diversity is a no-brainer, and you must do everything within your power to ensure progress continues to be made and it remains a key corporate priority.

7.6 Create the perfect storm by considering more apprentices to battle the war for talent

Kimberly Hepburn, Junior Quantity Surveyor, Transport for London

Kimberly Hepburn is a Degree Apprentice at Transport for London (TfL). Having entered the industry during times when apprenticeships were not fashionable, she has made it her personal mission to raise the prestige of apprenticeships and advocate this route as a worthy Plan A for those entering construction and property. In doing so, she received the first apprentice position on TfL's Graduate and Apprentice Board and the inaugural Apprentice of the Year title from the Royal Institution of Chartered Surveyors.

My name is Kimberly Hepburn, and I am product of a perfect storm of support, my discomfort zone and choice.

So what led me, granddaughter of the Windrush generation, into what is considered one of the corporate world's most white, male-dominated professions? Well, the seed was planted by my school's refurbishment, nourished with my parents' support, and let bloom by a CEO who believed in apprenticeships.

My social characteristics imply I shouldn't be in this field. I am a young, black, state-educated woman who was born in a working-class area and is a second-generation immigrant of Jamaican heritage. Moreover, when I chose to pursue an apprenticeship, they were not usually considered an option for "academic" students like me – they were seen as substandard – and all of my peers took the traditional route to university. Nevertheless, when they were graduating, I had become the inaugural Royal Institution of Chartered Surveyors Apprentice of the Year in 2017. Having worked on exciting projects during my apprenticeship at Transport for London, such as the upgrade of Victoria Tube station, I was, and still am, a passionate advocate for apprenticeships – mine gives me the opportunity to make a difference to people's lives every day.

My story is a perfect storm – but it shouldn't be. I entered this profession with something to prove, and I am determined to inspire the next generation. Today's task is encouraging CEOs to take on more apprentices.

What they don't tell you about apprenticeships

The brilliance of apprenticeships is often dominated by the widespread "earning while you learn" mantra. Here are three often overlooked reasons why those looking to enter real estate should consider an apprenticeship:

1   **Working alongside role models** – You gain motivation from working with colleagues who role-model what excellence looks like as well as being given the determination to achieve by seeing your future roles first-hand.
2   **Global employability prospects** – Many apprenticeships in real estate offer a fast-track route into a successful career, fully sponsoring and supporting your journey towards professional status, allowing you to apply your expertise to a global market.
3   **Educated by the industry** – Academia is just one part of the authentic education that apprenticeships provide. Apprentices get accustomed to 360° continuous learning and development, gained from their employer, training provider, professional body and endless development opportunities.

The apprentice who takes full advantage of these opportunities inherently becomes very knowledgeable about past practice, as well as the current and future needs of the industry – it's no wonder we are considered the "leaders of tomorrow". It's clear that apprenticeships are a no-brainer career investment with outstanding returns. But what's in it for you?

Why you cannot afford not to have apprentices

We know that diversity and inclusion isn't merely a hot topic. It makes business sense, and is particularly critical to a people industry like property. Those who overlook the importance of D&I run the risk of dampening their growth or driving their company to a premature decline. The best way to make your workforce more diverse is to ensure you have a variety of talent joining your industry, such as through apprenticeships. There will be a knock-on effect as they progress and make their way into leadership roles. The workforce will then reflect and better serve your customers. Apprenticeships are the secret key to unlocking your D&I strategy. What can a CEO learn from an apprentice?

It's likely that the apprentice in the room will have the least experience, but they can offer the freshest perspective. Aside from innovation and entrepreneurship, having apprentices as the foundation of your business allows CEOs to stay attuned with the desires and ambitions of the leaders of the future. The further removed you are from the next generation, the more likely you are to have untapped potential that would otherwise better equip you to respond to the needs of today and tomorrow.

Message to the CEO

Consider the impact of unconscious bias. Discriminate not between people or things, but between which thoughts you allow to evolve enough to get out of your mind.

"Are you SURE you want to let those thoughts out—aren't they biased?"

For some, this will be a radical mind shift, for others it will be just the invitation needed to make a real difference.

7.7 Look at the person as a whole, what drives them, and embrace it

Amy Leader, Associate Project Manager, Oxbury Chartered Surveyors

Amy Leader is a Chartered Surveyor working as an Associate Project Manager for Oxbury Chartered Surveyors in Cambridge. She is also a member of the RICS Governing Council and a proud mother of two children and two stepchildren. She enjoys a challenge, and is currently raising funds and training to cycle from London to Paris.

Twelve years ago my life was somewhat different. I was working evenings as a care assistant, looking after two young children during the day and questioning if this was it. I knew I wanted more out of life, I had always had a passion for the built environment, working my way up the property ladder by doing up houses and selling them on, and after much deliberation and a bit of coaxing, I plucked up the courage to go to university on a part-time basis to study Building Surveying. I was incredibly nervous about taking this on as I was diagnosed with dyslexia as a teen and struggled academically at school.

Within six months of starting my degree I became a single mum of two under-5s and remained a single mum for the duration of my degree. While working full-time as a trainee building surveyor, I was made redundant and informed verbally that this was in part as I was not flexible enough because I was a mother. I have watched others overtake me as I was passed up for promotion time and again as I continued to try and strike a balance between work, kids and my commitment to my profession.

On the flip side, I am proud to say that I managed to achieve a first-class degree and soon went on to become Chartered. I have been Local and National Chair of RICS Matrics, shortlisted for the Young Surveyor of the Year Awards, supported a number of students through their membership assessment, carried out numerous school visits, raised just short of £10,000 for the charity CRASH and was voted onto the RICS Governing Council.

So what drives me? I am passionate about making a difference, the challenge to grow as a person and to keep learning, whether that is by exceeding a client's expectations by delivering a project on time, on budget and to the delight of the end users, taking the opportunity to provide added value to the community I work in, stepping up and sitting on a board or standing at the finish line cheering my kids on at sports day.

However, this does mean that I, like many other experienced, driven and capable women, am missing opportunities for promotion because we do not put ourselves forward in fear that we will be asked to sacrifice too much time with our family, or that we are perceived as not being dedicated enough to the job. And yet there is so much evidence proving that this is a huge loss to business. A recent study by EY found that part-time working mothers are far more productive than their full-time colleagues, indicating that while we have less time, we make better use of it. In addition, *The Sunday Times* found that children of working mums do better at school, so it is not only good for business, it is good for the next generation.

With the rise in number of working mothers, I believe that this is progressively becoming a challenge for parents, and not just women. Fathers are taking on more of a role with childcare, so this is becoming just as much of an issue for men as it is for women, with fathers reluctant to ask for flexibility for fear of missing out on a promotion. I know from my own experience that my partner has missed out on opportunities, being advised that they did not feel he would have the commitment as he had asked for flexibility to see his children.

I have been fortunate enough to have had many a supportive manager in the past who has given me flexibility, safe in the knowledge that my job will be done. However, this has been on the understanding of "don't tell the boss", given the reasoning that if it was common knowledge, everyone would want to do it – like that's a bad thing?

My path so far has allowed me to meet so many inspiring people, to grow as a person, to be an engaged mother, to give back to my community and my profession as well as the next generation. I am lucky enough now to work for an engaging and supportive employer that encourages me to be my best. Believe me, all this really makes a difference, as I love my job and what I do.

Message to the CEO

So my challenge to you as a CEO is to look at the person as a whole, what drives them, and embrace it. Find the right person who will do the job in the way that works for the business and for them, and you will have a far more diverse, passionate, fulfilled, loyal, happy workforce.

"Hey—look at me as a whole, not a bunch of weird anthropomorphized pieces!"

7.8 Set the right culture, engender a spirit of kindness, and encourage more women to come forward

Pinky Lilani CBE DL, founder and Chair of the Women of the Future Awards and Asia Women of Achievement Awards

Pinky Lilani is a food guru, author, motivational speaker and internationally acclaimed champion for women. She is the force behind the annual Women of the Future Awards, the Asian Women of Achievement Awards, The Ambassadors Programme, The Women of the Future Network, The Women of the Future Summit and the Global Empowerment Award. She is an associate fellow of the Said Business School, Oxford, and Patron of DIL, The Westminster Society and Frank Water. She is an Ambassador for the Tiffany Circle of the Red Cross and on the Board of Trustees of the Royal Commonwealth Society.

Pinky is also a member of the board of Global Diversity Practice and the Court of Brunel University. Pinky makes regular media appearances and is a keynote speaker at international conferences and at educational institutions, including the Judge Business School, Cambridge and the Said Business School, Oxford. Winner of several awards, she was listed as one of the 100 most powerful women in the UK by BBC Radio 4 *Women's Hour*.

How important are the people and quality in your organisation? As a provider of professional services, my belief is that your people will be key alongside delivering a quality service to your clients.

If people are at the heart of your organisation, attracting the best, top-talent individuals will be paramount. In order to do this, you need to attract from the widest, best possible pool, from both genders. But with so few women entering the built environment profession, how can you embrace, attract, retain and develop the female talent to, and in, your organisation.

My name is Pinky Lilani CBE DL. My own diversity journey began in India, where I was one of the few Muslim girls in an Irish Catholic convent school. I met my husband in India, married him in three weeks and came to the UK – he had done no due diligence whatsoever, and thought I was an accomplished cook, but I had never been inside the kitchen as we had a fantastic chef in our home. Food and meeting people have always been my passion, I learnt to cook so that I could invite people over. I published two cookery books – *Spice Magic* and *Coriander Makes the Difference* – and marketed them by visiting corporates and book stores armed with my wok and box of spices. The aroma of fresh garlic and coriander in the most unlikely setting opened a huge market of opportunities. My belief in the importance of recognition for success led to setting up the Asian Women of Achievement Awards and the Women of the Future Programme. Kindness and collaboration are at the heart of all I do.

It was recognising that many of the exceptional women were often still in the minority and often lacked the confidence to self-promote themselves that led me to developing my now other passion by founding the Women of the Future Awards and Network. The network is for high-potential and high-achieving women, and provides an opportunity for talented women to come together, share experiences and build business relationships. The Awards provide a platform for female talent in the UK and Asia to be recognised and celebrated.

Why women? Because I am passionate about the outstanding skills and talents that I have seen in women that flourish when they become part of a network enabling more talent to grow. I recognise in others maybe what I saw in myself: the need for a supportive network and environment that enabled me to be the best that I can be.

In 2014 I was delighted that RICS helped develop and sponsor the first Women of the Future Award category for Real Estate, Construction and Infrastructure. Previously we had many categories covering a broad spectrum of disciplines from sports, media to humanitarian and business, but there was nothing that recognised women aged under 35 in this sector. In the first year of the awards the number of entrants for this category was higher than any other, and I am delighted to see that this still continues to be the case today. The network that the winners and runners-up have come to form is supportive and engaging with business leaders in the sector to help move the dial.

My reflections on the real estate, construction and infrastructure sector is that there is still such a low percentage of women entering and remaining in the sector compared to others. Yet it is a sector that desperately needs to attract top talent. Gender diversity is at the heart of solving this. What I have observed is that those women who do enter are incredibly passionate, articulate and committed leaders.

Message to the CEO

My message to you as a CEO in the built environment would be to say it is important to recognise first how important gender diversity is in the workplace. Helping encourage more women to enter the sector, and supporting those that do, is so important, especially for this sector where the percentages remain so low. This is not a case of supporting women over men, but providing a workplace environment where the differences each bring are recognised, nurtured and celebrated.

"A rising tide lifts all gender boats."

Of course women think differently to men, and having greater diversity at all levels in your organisations brings fresh perspectives, is likely to mirror your client's expectations, but will also deliver more valuable talent.

I have seen some of that talent bubble up through the Women of the Future Awards, and am constantly surprised by it – the passion these women have for the sector and how proud they are to be a part of it. They are all clearly great at what they do too.

Encouraging more women unquestionably makes business sense and will invariably help ultimately in the gender pay gap balance, but perhaps most importantly, doing more to attract and encourage gender diversity helps in opening opportunities for bringing in more top talent.

So why not look at what your organisation can do to bring more women in to the organisation at all levels – setting up policies like 50% male/female recruitment at graduate level, for example? Looking at wider inclusive policies such as offering increased workplace flexibility all helps, and providing support for these women to come together to learn, grow and develop will further fuel their commitment, as well as help retain them mid-career.

Set the right culture. Engender a spirit of kindness. Train your managers. Seek out the talent in the organisation and promote it – but proactively look for and encourage more women to come forward.

The paybacks to organisations such as yours are immeasurable in providing a greater balance and enabling a better working environment for all.

7.9 Look for the top talent – regardless of whether that person has a disability

Antonio Llano Batarrita, Llano Realtors S.L.

Antonio Llano Batarrita has been working in the real estate industry since 1990. In 1997 Antonio qualified as an Agente Propiedad Immobiliaria (API), an official Spanish real estate agent, and more recently as a chartered surveyor. He is a teacher, consultant, valuer and broker, and runs a small family-owned real estate company which was founded in 1976 in the Basque Country region in Spain.

My story is about not letting a disability get in the way of life and job fulfilment. I am a real estate residential agent specialising in valuations, market studies, land and property bankruptcies, and advising attorneys. I also teach courses and mentor real estate professionals. And I am deaf.

Deafness is a communication disability – the barrier is there, but is invisible. I am healthy, I go to the gym, run, swim, play football, but social relationships dwindle when somebody is deaf. There is no real physical barrier to overcome, but effective communication is the major challenge. Since I became deaf, I was forced to change the way I communicate with people, family, friends, partners, co-workers and customers. It is a two-way communication issue: if you don't understand what others are saying, you cannot answer them, so participating in an ordinary conversation is usually almost impossible, and hearing aids can help, but not really much!

Things are more difficult, but they are not impossible. The main goal as a deaf real estate agent is to control your communication. As misunderstandings pursue you, controlling conversation environments and ensuring and confirming dates and agreements in writing is already an old habit.

I know that there are things that I cannot do, and so I don't do them. I have never felt "disabled" because I have adapted myself to my situation. So, for example, I now specialise in more technical products that do not need full and permanent attention to what the customer is saying.

So how have I adapted?

I always encourage personal meetings, as it is the only way I have to communicate with people at work. In some cases it is bothersome and more time-consuming, but I accept there is no other way, and at least that way the meetings are under my control. I always choose places that have no background noise, and I never use big meeting tables or talk with more than three or four people at a time. So I always, always, always try to control the environment where meetings occur. I even attend conferences and conventions. I usually don't understand anything – but my real and fulfilling objective is to meet people and grow my business contacts.

The cellular phone is a modern and awesome tool, and most real estate agents use it extensively during for their business. In my case, electronic voice recognition software is a real challenge. I have my phone directly connected to the earphones and I only use it in order to set appointments, and usually write emails to confirm them. Not only does it give me the confidence that nothing can go wrong, but it is also good risk management.

And that's all! If you are disabled, you should be aware that nobody is almighty and there is always more than one way to do things. You have a disability, but you are not useless, so adapt, re-invent yourself and never give up!

Message to the CEO

My message to the CEO is that you don't need to make vast changes in your organisation to accommodate people with a hearing barrier. Encourage personal meetings so that deaf people can communicate at work. In some cases it may take a little more time to facilitate face-to-face, but allowing people to visit their customers or receive them at the workplace enables them to have much more control of the meeting. Creating workplace areas with little or no background noise and providing meeting tables for three or four people at the most not only allows sound volumes to be better controlled, but creates a much more inclusive environment.

"It's easy."

7.10 Look for the person who has the potential, not the finished product

Hashi Mohamed, Barrister at No5 Chambers and Broadcaster at the BBC

Hashi Mohamed joined The Honourable Society of Lincoln's Inn in 2010 and completed his pupillage at 39 Essex Street Chambers in 2012. Today, at No5 Chambers, Hashi practices in public law and commercial litigation-related cases, though his main area of focus is planning and environmental law. He has been consistently listed as one of the highest-rated planning barristers in England and Wales under the age of 35 in *Planning Magazine*'s annual Planning Legal Survey.

Principal perceptions

Life is really about perceptions. We'd like to think that it is in reality about substance, but it often *really* begins with perceptions – the perceptions we have built over the years through our own life experiences, the perceptions we learn from our own professional environments, the perceptions we share in our groups and circles. They can often have a profound effect on how we see each other and how we see others.

What you're about to read may well change your perceptions, but that will depend on how aware you *really* are about them already and how they impact your thinking, your point of view, your way of seeing the world. Let me begin by introducing myself.

My name is Hashi Mohamed. I am a barrister, and I also present documentaries on the BBC and am passionate about tackling the inequalities which exist in our country, in our communities and the world generally. That's me, today – but it was not always like this. Meeting me may lead you to believe, indeed perceive, that I have led a seamless and easy life, that I have had all the trappings of privilege and circumstance. But that would not be entirely correct. I came to the UK as a young, unaccompanied child refugee from Kenya, without either of my parents, having helped to bury my father at the age of 9.

My mother gave birth to 12 children, and we have been scattered around the world due to many seeking asylum in different places, including the USA and Canada. When we came to the UK, we could hardly speak English and settled in difficult neighbourhoods with poor-performing schools. Raised by my grandmother and extended families, we were not living in the most settled and stable of circumstances, raised predominantly on state benefits.

Evolving perceptions

Why does this all matter? Well, against the odds I find myself today working in one of the most elite professions, and it would be a great tragedy were I not to give back in a certain way. It begins with understanding better what's happened to me to date. My own story is made of a combination of extreme hard work, the right people believing in me at the right time, and some extraordinary luck. One of the ways I seek to give back is to find ways of changing perceptions. This is why I have taken the opportunity to address you here.

I address you, as someone working in the property and planning law world, with a view to encouraging you to think more widely about how you perceive the talent pool from which you recruit, to think differently about those who may not have the "normal" type of background from which you normally recruit, to think strategically about all the talents you're missing out on because of a certain group think, to think more carefully about how your own prejudices and stereotypes affect your perceptions. Social exclusion can affect different people in different ways, but can also be profoundly damaging for those who face a "compound" combination of discrimination, particularly those from BAME backgrounds.

I do not think there's anything exceptional about my starting point, but the desire, determination and hunger to succeed are things I gained from those around me. I truly believe that your starting point must never determine what you can do or how far you wish to go. I am far removed from the typical starting point for the majority of those in my profession, but this is not something which destabilised me or indeed made me feel inferior.

However, it is true to acknowledge that the majority of those with whom I share a background either ended up on drugs, behind bars or just falling into the wrong crowds at critical moments.

My story is that of a transformational change – a transformational change that I believe is within all of us. But in order for it to flourish, it required support, imagination, encouragement and scholarships – but also deep determination, hard work and an innate desire to make a difference. But also, of course, some luck. In addition to my professional commitments, I seek to build the right environments to allow more mentoring, coaching and supporting others to have the courage to fulfil that which is within them.

Message to the CEO

My message to the CEO would be: look for the unseen, seek out the rough diamond, look for the potential instead of being seduced by the polish, don't be quick to judge based on how someone looks, sounds or what school they attended. Look for the person who has the potential, not the finished product, the person who is full of talent and energy, and has the furthest yet to grow – I promise they will be worth it in the end. Do not be afraid to take risks.

"Woah, that is one rough diamond—could he become one of us?"

Give people a chance to work hard and prove themselves; support those in apprenticeship programmes who look and feel different to your own and organisational norms, not out of pity, but to find new and creative ways of working and discovering – you may just be surprised with the payback you get.

Courage, confidence and imagination are not just necessary for me the individual, but for you the CEO to think about the opportunities you can give and receive.

Make your moment count, and give someone unusual, but also special, a chance to make a difference in your organisation, to help them realise their full potential – and your own organisation the chance to find new heights.

It will be worth it.

7.11 A diverse workforce is not only good for your image, but is good for your profit margins!

Ali Parsa, Professor of Real Estate and Land Management, Royal Agricultural University

Ali Parsa is currently Professor of Real Estate and Land Management at the Royal Agricultural University (RAU), UK. He joined the RAU in September 2012. Ali was Professor in Real Estate at the University of Salford between 2010 and 2012, and prior to that held the Rotary International Chair in Urban Planning and Business Development in the University of Ulster from 2007 to 2010.

His research is multi-disciplinary in nature, but focuses primarily on emerging real estate markets, Shariah-compliant real estate investment, and the interface between urban planning and real estate markets, globalisation and urban development. Recent projects include "Viability and the Planning System: The Relationship between Economic Viability Testing, Land Values and Affordable Housing in London" for a consortium of 13 London boroughs, real estate strategies for tall buildings, environmental impacts of real estate and construction, urban land management, and housing policy and housing market analysis conducted for private and public sector clients in Europe, the USA, Asia and the Middle East. Ali was recently involved with an EU-funded project on Commercial Local Urban District (CLUDs) – a comparative study of urban regeneration in the UK, Italy, Finland and USA involving a team of eight professors and 18 researchers in Italy, Finland, the UK and USA.

We live in interesting times when much of what seemed to be normal does not appear so normal and when traditional and not so traditional institutions are being challenged daily. As someone who has been involved in global real estate education for nearly 30 years, I have witnessed tremendous change both in the real estate sector and the profession.

I was born in Iran, and came to the United Kingdom in pursuit of higher education in my late teen years in 1977. I am neither a political nor economic refugee. I am part of that growing and highly mobile global workforce that has been shaping the globalisation of real estate. As a senior academic, I often find myself devoting much time to students who seek advice about their future and what path to follow in the real estate industry.

What does not make the headlines is how the real estate sector has had to change in the way it has treated gender and an increasing diversified workforce. I recall vividly when a number of my "British White Female" students were having a lively conversation about their work experiences in the office, which included being treated as a tea lady at break times, with their male line managers demanding they make tea or coffee. Others were complaining of not being taken seriously and being given routine office work such as photocopying and sending faxes. That was in the late 1990s and early 2000s. I can also tell you about the frustration of my minority students who had passed their BSc and MSc courses with top marks, but were still unable to find graduate training opportunities within the mainstream private commercial and residential property companies. Either by design or coincidence, the majority ended up as rent collection officers in the public residential sector (again in the not so distant past). It was particularly intriguing visiting many of the leading property firms over the years and not coming across often a single staff member from a minority background. This was not confined to the UK. I can also recall entertaining visiting academics from a couple of European countries who were surprised to find "black students" studying real estate – they used to ask about this in amazement!

Real estate is a major contributor to wealth-generation, and therefore is influenced by cultural and societal values. The globalisation of real estate and the insatiable appetite for property as an asset class by international investors combined with changing legislation relating to diversity and inclusion are the main drivers for change. A diverse workforce is prerequisite to success of any business.

Message to the CEO

The message I want to give a CEO or senior leader in this sector is that your workforce are your human capital and the most valuable resource. They are the source of innovation, talent and entrepreneurship. A diverse workforce is not only good for your image, but is good for your profit margins! How talented are you in utilising your best resource?

"I've struck capital—human capital! Fleshy gold!"

7.12 Workplace wellness begins with just talking, wellbeing and introducing practical solutions

Gian Power ACA, founder and CEO of TLC Lions and Unwind London

Gian Power set up his first business aged 13, and later worked at Deutsche Bank and PwC and witnessed first-hand some of the diversity and inclusion issues that need to be tackled in the corporate world.

In 2015, aged 23, an unexpected family tragedy changed Gian's life for ever. His experiences and inner resilience sparked his passion for prioritising wellbeing and inclusion in the workplace. Gian is the founder of TLC Lions and Unwind London, with clients including GSK, Sony Pictures and Lloyd's of London. His work has been recognised globally by *Business Insider* and *The Independent*, and will be the subject of a BBC1 documentary in 2019.

Gian is the founder of TLC Lions, a collective of ordinary people who share extraordinary stories to ignite emotion in organisations. In these powerful, personal and emotive sessions, the Lions (speakers) share stories from the heart that build empathy and understanding in the audience. Gian later founded Unwind London, Europe's first surround sound meditation experience, welcoming organisations to the very thing that saved Gian – meditation.

Gian firmly believes that wellbeing is no longer a "nice to have", it's a business imperative and has a direct impact on employee happiness and productivity. As an international speaker himself, his message is now global and is supported by individuals at Harvard and Massachusetts Institute of Technology exploring the future of work.

From appearances, I am your average 26-year-old, professional and confident. I've been told my bold nature and drive projects strength to those around me. That projection, as many senior executives have expressed to me, is what they feel they have to show in the workplace just to be taken seriously, afraid to show any vulnerabilities or weaknesses around boardrooms or in front of their employees.

Four years ago I wouldn't have understood the struggle to project strength while crumbling inside, the lasting damage a workplace culture such as this can do to someone, how it can affect their work, their productivity, how they, in turn, treat their staff and colleagues, then subsequently their home life and ultimately their mental health. I didn't understand the power of sincerely asking a colleague how they were, or authentically caring about what is happening to them beyond their work. Just checking in with someone on an emotional level can be so empowering for them and your relationship, it can turn their day around, make them happier just to have been heard and make them take on their work with a renewed zeal.

Four years ago I wouldn't have even contemplated workplace wellness as a factor in anybody's life. How wrong was I?

Everything changed on 8 May 2015. Some may recall it as the day of the general election in the United Kingdom; however, for me it was the day that changed my life for ever. It was on 8 May 2015 that my father was murdered.

Suddenly this "normal" graduate life turned into fighting an international investigation, endless court battles, media interviews, grief and trauma, all while trying to hold down my job at PwC and complete my ACA.

Three months later, it was time to return to work. I returned more self-aware than ever, with my emotional compass on high alert and taking my self-care very seriously. As a result, I soon realised that everyone has a story they would like to share if only we are willing to take the *time* and listen.

I still vividly remember the first day I experienced one of the most important tools in my self-care arsenal. The police had come to my office to speak to me about the investigation, the media were outside, and I had client work to be done. It was as surreal as it was overwhelming. Then a friend handed me a video and told me to take some time, find a quiet space and listen. This was my first introduction to meditation, and I've never looked back.

Those 10 minutes took me out of the intense situation that had been suffocating me, it calmed me, balanced my mind and allowed for some of the best decision-making to date. Ever since, it has become an integral part of my own self-care and my work. When two friends were hospitalised in 2018 due to workplace exhaustion, enough was enough. Unwind London was born.

In less than four months we introduced thousands of employees to the power of meditation, transforming offices across the city into candlelit experiences tailored to a corporate audience. It's proven that looking after employees and their wellbeing boosts productivity, job satisfaction and staff retention.

Wellbeing is a priority. Self-care is everything. The better we feel, the more productive we are. We might not change the long hours, late nights and always-connected culture we see today, but as leaders, we must do more. Companies that don't realise the value of wellbeing at work will inevitably lose top talent to those that do – or worse, because they are too unwell to continue.

Message to the CEO

Consider your top employees right now: are they looking after themselves? Are they working at 100%? Would you even know if they weren't?

Take some time to ask. Show them some feeling and compassion in their own struggles. Invest some of their working hours into their wellbeing. Provide easy-entry solutions like meditation to begin the journey. It doesn't take much to transform those employees, their work and the overall culture into something that will pay dividends for everyone for years to come.

"Maybe that's too much meditating."

7.13 Flexibility is key; flexible working conditions will lead to a creative working environment

Sanett Uys, Managing Director, Serendipityremix

 Sanett Uys has worked in the commercial property industry for over 20 years. Previous employers include Rode & Associates, Broll Property Group, Colliers International, and Excellerate Valuations and Advisory Services.

She is the founding member and Managing Director of Serendipityremix, a private property consultancy which started trading in February 2016. Sanett's academic qualifications include an MBA from the University of Stellenbosch.

She is Chair of the RICS Sub-Saharan Africa Market Advisory Panel, is a former member of the National Council for the South African Property Owners Association (SAPOA), a former committee member of the South Africa Council of Shopping Centres (SACSC), and the previous National Chair of the Women's Property Network.

In the previous 20 years of my career in the real estate industry, I've had the privilege of working with a variety of people who helped to shape my career, picked me up when times were tough and guided me to find my path, even helping me correct course a time or two. Without all of them, I would not be the person I am today.

Who am I? My name is Sanett Uys, I am a wife, mother of three beautiful daughters, community activist and the owner-operator of a small specialist property services company in South Africa. I grew up in a middle-income family, and due to my father's retrenchment and my not being sure what I wanted to study, I took a year's sabbatical and started working. A year became three years, and then I got an opportunity to work in a property research and valuations firm, and the rest, as they say, is history.

I started studying part-time, and in seven years got three degrees. In these seven years I learnt a lot about time management, working flexi-time, multi-tasking, understanding people, and most of all, I learnt to endure.

My story is not unique, and there are so many similar stories I can share about colleagues in the industry who walked a similar path. The themes of our stories are comparable, as it is all based on hard work, dedication and putting in the long hours required to be successful. Success achieved! But then we move into the next phase of our lives, where we have to balance our work and family lives. To most women, this is very challenging. So many unknowns and questions are constantly going through our minds: are we doing the right thing, are we putting our families first? How do we get to a work–life balance?

Message to the CEO

There are no superwomen out there: from interns to board level, we all get the same 24 hours every day, and we are just normal human beings working hard to keep everything together. Trust us, it is not always easy. The challenges are real, and so are the feelings of guilt and inadequacy – guilt for not always being there for our families, especially our children, guilt for leaving the office before some of our colleagues to see to our families, and most of all, feeling guilty because we love what we do so much that sometimes we don't want to go home.

"Don't be a Superwoman—give yourself a break!"

Some women have an even bigger challenge: being a single working parent with a limited support system. This working single parent is everything and all things to the family they are raising. A single parent is one of the most amazing employees an organisation can have, as they are extremely loyal, masters at multi-tasking, and hardworking because their family's livelihood depends on their income. Make sure you look after these employees, give them what they need to excel in their work environment, and remember that there is no "one size fits all". Flexibility is key: flexible working conditions will lead to a creative working environment, and time is more important to a single parent than just pure income. Trust them to deliver the high-quality job you employed them to do. They will repay your understanding and support tenfold. Have some empathy, understand that it is not always easy, and sometimes a smile and a thank you is all they need. The results will be overwhelming, and the output will increase substantially. Trust me.

7.14 If you want to widen participation, you have to widen access

Ashley Wheaton, Principal, University College of Estate Management (UCEM)

Ashley Wheaton has spent over 25 years in the learning and education sector. At Microsoft, as Director of Global Learning Services, he was responsible for the education of 14,000 staff worldwide, and more latterly, in his role as Principal at UCEM, he oversees the higher education provision for 4,500 students studying worldwide for undergraduate and postgraduate degrees in the built environment. UCEM will celebrate its centenary year in 2019, and is recognised as the largest single provider of qualifications for built environment professionals worldwide.

He is deeply passionate about education and the wider skills agenda. This combines with his determination and commitment to use technology as the means by which education can be made not only better, but also more flexible – and critically, more accessible for all learners.

Ashley is also a Non-Executive Director at the Building Research Establishment (BRE), and Trustee at the Chartered Surveyors Training Trust (CSTT). He holds a BA in Economics from the University of London.

As a qualified economist, I like to look at most things through the lens of "supply and demand". My role as head of an institution with a clear purpose to generate the supply of qualified students with the skills necessary to enter careers and professions in the built environment is to ensure that the access for students to education is free from as many barriers as possible, and thereby through the provision of relevant academic programmes, present the widest, most diverse group of qualified graduates to industry.

At UCEM we stand firmly behind widening access. We don't just say it or talk about it, we do it. Our stated Core Purpose (a concise summary of the objectives in our Royal Charter) has six core tenets. These are a very precise and important representation of why UCEM exists, and they are reflected in everything that we do and the subsequent daily actions of all of our staff. The very first (and most important) tenet is to be "accessible". Our view is that if you want to widen participation, first you have to widen access.

To do so, you have to remove the barriers which prevent students from studying in the first place. You have to eliminate, or at least minimise as far as possible, the financial, social, geographical, physical and academic barriers to education, and thereby enable access to anyone who wishes to study for a relevant qualification and develop their career in the built environment. That's what we do at UCEM, and why I am so proud of the work that we do to support thousands of students in over 115 countries, many of whom would simply not have the opportunity to study if UCEM programmes weren't available. At UCEM, our learning model is specifically aimed at catering for all locations, backgrounds and needs. It's low-cost, and study can be done flexibly around work – benefiting those on lower incomes and supporting those with children and other responsibilities. The programmes can be completed from the comfort and security of your own home, potentially helping those with disabilities. There are also reasonable entry credentials and levels of final award, which can aid those without previous access to higher education.

Of course, access must be followed by meaningful outcomes, which assist industry in delivering on its ambitions, as well as enhancing its breadth and depth of participation. The UCEM Core Purpose goes on to state clearly that it is not only about accessibility, but also the provision of relevant and cost-effective education, which enables students to enhance their careers, increase professionalism and contribute to a better built environment – in other words, not just access for its own sake, but access with intent, real value and higher purpose.

It may go without saying, but the "demand" side must also be healthy. CEOs of organisations within the built environment need to put real substance behind their stated aims to diversify their employee populations. The rhetoric (good as it is) must be backed up with action. The sad fact is that the majority of the country's biggest companies still have no strong ethnic minority and mixed-gender presence. Despite government recommendations that no FTSE 100 board should be exclusively white by 2020, a recent report found that almost 58% have no ethnic minority main board presence.

While the built environment sector is making progress and the industry may appear more diverse – there are particularly more women, different ethnicities, various sexual orientations and people with disabilities within the new wave of apprentices – most of these people are just at the start of their careers. The challenge which CEOs need to face up to is that it will take a very long time for this younger generation of professionals to flow up into management, more senior roles and the boardroom – assuming even that they remain in the industry for that long. If the built environment simply waits for the lower end to move up and start having an influence, they'll be waiting 20 years or more for this to have the full impact, which isn't anywhere near quick enough.

Message to the CEO

My message to CEOs is simple – it's time to act now. As leaders in a sector with stated aims to be more diverse, you need to look at all of the options, at all levels of experience. You need to be open to educating and recruiting a diverse range of qualified people at all career levels, from top to bottom: there are plenty of mature candidates with experience and senior-level skills who could make a difference in the built environment, as well as the younger, less experienced generations who are just starting out. It's our joint responsibility to make our sector an attractive place to be a part of and ensure that everyone who wants to be part of it can be. At UCEM, we're committed to building the supply, and we need industry to be open to enhancing and developing the demand.

"Now that's a career ladder we can get behind!"

7.15 D&I is more than just another corporate buzzword

Sue Willcock, Director, Chaseville Consulting Ltd

Sue Willcock has worked in and around the construction and real estate sector for almost 30 years, having originally trained as a chartered surveyor.

In the last 15 years she has focused on people development and business change and has held partner and director positions.

She is author of the Amazon best-seller *Help, I'm a Manager*, and has recently released her new book, *Help, I'm Starting Work*. Both are niche books focusing on helping people make transitions during their careers in professional services.

If you asked most CEOs about the business case for diversity and inclusion, I believe most could come up with an intuitive response that would make sense. Indeed, there would be lots of nodding heads from most people if we reflect on how diverse teams we've worked with have often challenged our thinking, made idea-generation richer and perhaps questioned our own assumptions when coming up with solutions.

Despite this, it's still too easy to think of diversity and inclusion as another corporate buzzword, another target, another KPI. I believe it's much more powerful to go back to basics. What does it really mean to individuals and the behaviours we need to develop in organisations?

Message to the CEO

Here are a few thoughts as a woman in the sector from a socio-economic background where had I not been supported by people like you, I would have had a very different life indeed. I was lucky. The bad stuff (the assumptions, the sexism, the very "middle-classness") was far outweighed by some really great stuff. Almost 30 years on, having been a partner in a Top 5 UK consultancy, having led people development programmes in the sector and having written books for managers and newcomers in professional services, here's what worked and what I believe we need to see more of to reflect the society the real estate sector serves:

- When planning recruitment, pause to think about the likely outcome of your approach. Will asking current employees to recruit via their networks promote diversity? Putting an advert in one social media stream may get you a different response to another. Where and how do you source graduates? Do you have an apprentice scheme? Do you link with your community leaders to support the attraction of diverse candidates?
- Seek some help to understand unconscious bias that may exist in your organisation. Some locations will naturally be more diverse than others, and across your business you may need to help people recognise, understand and embrace difference.
- Cultivate a culture of being the "best version of yourself" – for everyone. The best places I've worked with are those where it's OK for anyone to say "I'm leaving early to pick up my son today" or to wear clothes that are role-appropriate, not an "unwritten uniform" that just does not work for many (indeed, sometimes it even intimidates or alienates).
- Avoid "one size fits all". Yes, we often need policies and governance to make things work effectively. But there are rules that by their nature can detract from the creation of a diverse workforce. Working hours, dress codes, the ability to work from home or remotely – all of these have a massive impact on your ability to attract, retain and engage.
- Create an open culture of questions and listening to understand. Often all it takes is for a confident person to ask someone a personal question to create a moment of learning, connection and embracing difference. If we can encourage confident conversations and genuine learning, then this fosters understanding and inclusivity.

I've seen some glaring examples of non-diverse and uninclusive design recently. Prayer rooms at the far end of a corridor added as an afterthought. A lack of thought for nursing mothers in public buildings where they are core customers. Place-making where "shared space" schemes, where buses and pedestrians mix, scare parents of young children and the elderly. We know we have the power to shape our built environment. For me, there is no better reason than this to ensure we have the right people around the table in our organisations to help create a diverse and inclusive society.

It's important to have the right people around the table.

8 Be the change that you want to create

8.1 Introduction

Real estate and construction have not been traditionally diverse or inclusive in terms of workforce. While much has started to change in recent years, it is interesting to note that the number one concern emerging from employers based on RICS research was tackling the war for talent.

As a service provider, the number one concern is, quite understandably, having the best people in your business to service your clients' needs. Getting those people is difficult when across the board there is a strong demand for top talent, not just in the sector, but also in science, technology, engineering and maths and all other sectors.

As the CEO, you set the vision for the organisation. Your role as a leader in tackling D&I as a mechanism to attract, develop and retain the best people in your organisation, as well as setting the cultural tone for the organisation with your clients and suppliers, is critical.

Your people, your employees, are your greatest asset. They need to be treated fairly, whoever they are and wherever they are in the world. It is important to develop a single vision for the organisation that can be adapted depending on the global context.

8.2 Visionary leadership

D&I is not necessarily a subject that will sit well in all parts of the business and with all people. It is certainly restrictive in terms of the extent to which the D&I protected characteristics can be discussed or implemented across the globe. It is important to recognise that a global vision has to take into account national practicalities, since in some countries it is illegal even to discuss or disclose these characteristics.

The role of the CEO is to be able to think ahead and plan for the future, taking on board political, social, economic and technological changes relating to market opportunities. Setting the vision for what the organisation needs to do on D&I is important for many business reasons – financial performance, legal considerations, market impact and the people dimension – all of which are detailed in Chapter 6.

What is important is people. Investing in your employees to attract, develop and retain them is essentially about helping them to become better employees. Creating the right culture, environment and workplace that enables them to bring themselves to work every day and to feel comfortable in the workplace will engender greater loyalty and a work ethic, as employees will want to give their best.

There is a greater recognition now among organisations in our sector that D&I is critical for the future success of the organisation – for example, it is why 177 companies (of all sizes) have now signed up to the RICS Inclusive Employer Quality Mark, representing over 300,000 employees.

A commitment to D&I starts with you, the CEO, at the top of the organisation, and it is certainly in demand by those you are taking on as juniors (apprentices or graduates) at the other end. A lot of leadership is required in supporting the middle layer of the organisation – middle and line managers, who need to be supported in embracing, adopting and implementing the change. Remember that sometimes it can be the smallest things that make the biggest difference.

"Companies are like sandwiches—the middle matters."

Wherever you sit in your organisation's hierarchy, it is important that, as a leader, you are inspiring, empowering and competent, but to be successful on the

D&I agenda, you need to also be authentic, empathetic and compassionate. Show some vulnerability, and listen well to those around you so that you embrace the plurality of voices and perspectives. People need to be heard and believe they belong – do not shoot down, but lift up and show that the organisation cares.

8.3 Writing the vision

The vision essentially documents a statement of what you envisage your business to be at some stage in the foreseeable future, based on your organisational goals and aspirations. Having a clear vision will give your business a clear focus, as well as preventing you from heading in the wrong direction.

There are strong, changing market dynamics at play – which makes the setting of the vision even more important.

A good vision statement should be short, simple, clear, and specific to your business and organisation, but it should also have within it a robust level of ambition.

Some things to consider:

1   **What do your people want** from the organisation now, and what are they likely to want in the future?
2   **Create the right conditions and environment** so that employees can be in the best possible workplace.
3   **Put yourself in your client's shoes.** What sort of organisation do you need to create that ensures your values and those of your customers are aligned?
4   **Seek an independent assessment of your organisation** that will show you what you are doing and will give you guidance about the gaps compared to the competition.
5   **Tap into the organisation to better understand what your people think and want.** Establish what would make the most difference to them. Use social and informal networks as well as more formal channels.
6   **Look at the competition** – both within the sector and beyond it – to really understand what leading practice is and learn from the experience of others.
7   **Use your own networks** to test your thinking and develop the vision.

It is important as a leader that you can convincingly communicate the vision. So remember that when it comes to D&I for your organisation, you are establishing something that will:

- be more likely to create better client outcomes;
- enhance your brand in the market;
- improve your reputation;
- achieve better returns for your shareholders;
- attract, develop and retain the top talent;
- create a better workplace for your employees;
- be more innovative in thinking and approaches.

Companies in the top quartile for racial and ethnic diversity are 35% more likely to have financial returns above their respective national industry medians.[1]

Remember that D&I is the solution, and is a matter of organisational performance, not an issue or problem to solve.

"Put away that, abacus, Farley—let's solve this problem together."

8.4 Walking the talk

The culture of the organisation forms its DNA. The organisation's approach to D&I is an important part of this culture. This starts with you, as the CEO. Think about what you can personally embrace that says you really want to walk the talk when it comes to D&I:

- Attend, speak at, and support networking events.
- Make a personal pledge that you will only speak at or attend events that are diverse and inclusive.
- Call out behaviours that are not inclusive of the vision you are setting.
- Recognise and reward the right behaviours.
- Ensure D&I is part of the agenda for every board meeting, to show that you are serious about it.
- Put in place key training and leadership development for all the leaders in your business on key D&I issues, such as unconscious bias, interviewing and performance management, and attend these sessions yourself.
- Listen to and engage with your employees at all levels, as well as your clients.
- Find a reverse mentor and engage with the next generation.

- Set up junior or shadow boards to get the next generation of leaders more engaged with the current ones.
- Communicate what you and the organisation are doing.

8.5 Tackling the agenda

Tackling the agenda head-on is now key. It is time to stop focusing on policy, and instead to focus on closing the implementation gap – by taking action. What you and your organisation need to do in relation to tackling this agenda and winning the war for talent, as outlined in earlier chapters, can be summarised as follows against the new four principles implemented as part of the IEQM.

"I'm going to tackle this agenda and win the war for talent."

Leadership

Diversity and inclusion is not about ticking boxes, you have to believe it and implement real change, starting with education. Go for industry accreditation, but go beyond that and look to drive a paradigm shift in culture and behaviour. Do not underestimate the commercial imperative of having a diverse and collaborative workforce. From my own experience, I know that if you get this right, you will be a more successful business, where people will thrive and will want to stay. The barriers to achieving this are challenging and there are no quick fixes, but the rewards and the ability to create something special are off the scale.

Ciaran Bird, UK Managing Director,
CBRE Ltd

(i) Show strong leadership

Look for the unseen, seek out the rough diamond, look for the potential instead of being seduced by the polish, don't be quick to judge based on how someone looks, sounds or what school they attended. Look for the person who has the potential, not the finished product, the person who is full of talent and energy, and has the furthest yet to grow – I promise they will be worth it in the end. Do not be afraid to take risks.

Give people a chance to work hard and prove themselves; support those in apprenticeship programmes who look and feel different to your own and organisational norms, not out of pity, but to find new and creative ways of working and discovering – you may just be surprised with the payback you get.

Courage, confidence and imagination are not just necessary for me the individual, but for you the CEO to think about the opportunities you can give and receive.

Make your moment count, and give someone unusual, but also special, a chance to make a difference in your organisation, to help them realise their full potential – and your own organisation the chance to find new heights.

It will be worth it.

**Hashi Mohamed, Barrister at No5 Chambers
and broadcaster at the BBC**

- **The D&I agenda and vision needs to start at the most senior level within the organisation in order to be successful.**
- **Have a D&I policy for the organisation that shows a public commitment to the agenda, and embrace operating to a standard such as IEQM, or be assessed to a standard, such as the NES.**
- **Set the dial for D&I within your organisation,** whether as part of corporate social responsibility or beyond it into the culture you wish to create.
- **Set the vision and strategy for D&I within the organisation,** taking into account the organisational culture you wish to create.
- **Demonstrate your personal commitment to this agenda.** Take personal ownership, be visible and accountable for the agenda;
- **Consider your own attitude and that of your senior management team** towards how D&I business practices can be embedded into the organisational culture, and develop action plans.
- **Get the full commitment of the leaders in your business** that they too not only support this, but will endorse it with their teams and correct non-conformance.

- **Put in place training for your leadership team so that they understand and embrace the D&I agenda.**
- **Allocate time to develop and implement the strategy**, to give it focus and to ensure effective implementation.
- **Establish the organisational baseline** to have an honest understanding of where you are and to demonstrate progress over time.
- **Ensure D&I is embedded in all business decisions and is part of the overall business strategy.**
- **Engage key stakeholders and change ambassadors.** Consider:
  - actively engaging with the corporate network, specifically on the topic of D&I;
  - sponsoring events where D&I is a theme;
  - organising D&I-related events for stakeholders;
  - trying to win prizes or sponsoring awards;
  - enabling people in the organisation to speak about D&I at events;
  - publishing about the topic in relevant newspapers and magazines
- **Review governance and management** by putting in place new governance and management responsibilities or adjusting existing ones.
- **Monitor the changes through a D&I impact assessment** and understand the reasons for what is being presented.
- **Put in place an action plan to make the changes the organisation requires** with clear accountability and responsibility.
- **Consider setting targets or aspirational objectives for your organisation and the leaders of your business** to help move the dial on D&I.
- **Remember to be authentic – if you say it, do it.**

(ii) Be aware of some of the key challenges

- **Understand the global challenges of implementing D&I** and the legislative implications.
- **Gain an understanding of the protected characteristics and how these apply** for the countries in which your organisation and clients operate.
- **Know the boundaries and restrictions for data collection and inherent culture** on what can and cannot be collected (or said) in the countries in which your organisation operates.
- **Size matters, and will influence what you need to be considering** – the size, purpose or sector in which you operate and whether you are an international firm or an SME or anything in between will impact what you need to be doing and how you roll out D&I in your organisation.
- **Where resources are limited, consider how best to leverage what you have** in order to create an impact using technology and through collaboration.
- **Know and tackle the reporting and challenges of gender pay gap reporting for your organisation**, and look at ways in which you can reduce the gender pay gap.
- **Do not go it alone – always involve your HR and legal teams.**

(iii) Set the culture, language and behaviours

- **Understand and set the culture for the organisation** by being aware of the culture that exists and in which you operate as an organisation.
- **Support your employees in gaining broader cultural knowledge and experience** to diversify perspectives and challenge engrained thinking.
- **Consider and review the business language and policies for the organisation,** and make sure everyone can be included.
- **Alongside the leaders in your organisation, demonstrate the behaviours you wish to instil, and** make sure everyone knows and understands what is acceptable.
- **Be open-minded about national and regional cultures and the "why"** by considering hiring people that are different and from varied backgrounds to challenge the thinking and foster open-mindedness and tolerance.

Recruitment

The message I want to give a CEO or senior leader in this sector is that your workforce are your human capital and the most valuable resource. They are the source of innovation, talent and entrepreneurship. A diverse workforce is not only good for your image, but is good for your profit margins! How talented are you in utilising your best resource?

Ali Parsa, Professor of Real Estate and Land Management, Royal Agricultural University

(i) Support attracting the next generation

Consider the impact of unconscious bias. Discriminate not between people or things, but between which thoughts you allow to evolve enough to get out of your mind.

For some, this will be a radical mind shift, for others it will be just the invitation needed to make a real difference.

Kimberly Hepburn, Junior Quantity Surveyor, Transport for London

- **Get into schools and support schools' initiatives** to encourage young people to think about a career in the profession.
- **Educate the educators** – persuade schoolteachers and parents who help influence early career decisions that a career in property is a great opportunity.
- **Support an internship and apprentice programme** to help people from a wider background to enter the organisation.

- **Extend the graduate programme** to people with cognate and non-cognate degrees.
- **Help change gender perceptions by working closely with the next generation** to understand what will make a difference for them.

(ii) Recruit the best talent

> Look at the person as a whole, what drives them, and embrace it. Find the right person who will do the job in the way that works for the business and for them, and you will have a far more diverse, passionate, fulfilled, loyal, happy workforce.
>
> **Amy Leader, Associate Project Manager,**
> **Oxbury Chartered Surveyors**

- **Seek to engage and attract new people to the organisation** from underrepresented groups.
- **Put in place the leading practice recruitment methods** that embrace your organisation's vision and aspirations:
  - Consider D&I throughout the recruitment process.
  - Consider engaging and attracting new people to the industry from underrepresented groups.
  - Starting salary and parity across male and female recruits is a relevant consideration.
  - Apprenticeships and internships are effective ways of bringing new talent into the organisation.
  - Consider non-cognate or mature entry candidates or those from outside the sector, such as the armed forces, to diversify the intake.
- **Recruit, where possible, on the basis of a 50:50 gender balance,** particularly at graduate level.
- **Create a positive experience from the moment someone has contact with your organisation** so that no matter what the outcome, they feel the difference.

> My message to CEOs is simple – it's time to act now. As leaders in a sector with stated aims to be more diverse, you need to look at all of the options, at all levels of experience. You need to be open to educating and recruiting a diverse range of qualified people at all career levels, from top to bottom: there are plenty of mature candidates with experience and senior-level skills who could make a difference in the built
>
> *(continued)*

(continued)

environment, as well as the younger, less experienced generations who are just starting out. It's our joint responsibility to make our sector an attractive place to be a part of and ensure that everyone who wants to be part of it can be.

Ashley Wheaton, Principal, University College
of Estate Management

Culture

Look beyond the obvious and move towards ensuring that people are judged based on outputs, and not perceptions. My mission is to ensure that we attract, retain and enable the attainment of diverse talent, but without them having to outperform. It is important that the hurdles for promotion are the same for everyone.

Justin Carty, Senior Director,
CBRE Capital Advisors Ltd

(i) Create the right workplace environment

You don't need to make vast changes in your organisation to accommodate people with a hearing barrier. Encourage personal meetings so that deaf people can communicate at work. In some cases it may take a little more time to facilitate face-to-face, but allowing people to visit their customers or receive them at the workplace enables them to have much more control of the meeting. Creating workplace areas with little or no background noise and providing meeting tables for three or four people at the most not only allows sound volumes to be better controlled, but creates a much more inclusive environment.

Antonio Llano Batarrita, Llano
Realtors S.L.

- **Create an environment that is open, transparent and inclusive for all** such that people feel comfortable to be who they are and give their best.
- **Understand cultural differences in different countries.**
- **Consider the implications of the multigenerational workplace** by looking beyond the generation (age) or number of years of service, and look instead at the overall contribution.

- **Formulate an inclusive workplace for those with disabilities**, for example:

  - People with disabilities should be provided with clear information, in a format they can understand, about what support or changes to the workplace are available to them.
  - Consider how buildings, services and the workplace environment can be adapted so that they can access and use them safely and freely alongside their colleagues.
  - Think about what additional training and support may be required to help them do their job effectively in the work environment they choose.
  - Provide help and support before things get on top of them. Managers and colleagues need to be supported to understand what signs to look for.
  - Think about what technology and gadgets can be provided to help support people to be more independent and to work more effectively.
  - Flexible working allows people to not have to always be in the office if they need to work closer to home, or at home, from time to time.
  - Provide support to keep their job if they become disabled while they are employed by you.

- **Recognise the importance and key attributes of creating a family-friendly workplace,** such as:

  - understanding the importance of being able to see flexible working role models (both women and men) to support a healthy work–life balance – this is equally attractive to both women *and* men;
  - facilitating the challenge of caring for young families as well as parents and relatives who need care and attention;
  - allowing flexible working – this is a decisive element for organisations to increase their competitive edge, recruit high-potential individuals, retain qualified staff and maintain high engagement levels;
  - accepting that traditional working models are too focused on being physically present, and that this does not necessarily guarantee high engagement;
  - relying on result-based working models; irrespective of a home or office location, results are measurable on the basis of it is not *where you are* but *what you do* that counts – this needs to be fact-based and thereby completely transparent.

- **Have faith in employees working flexibly away from the office** by introducing a results-based working philosophy, rather than relying solely on counting on a physical presence in the office.
- **Make it an inclusive culture** where all staff engage with developing, delivering, monitoring and assessing D&I for the organisation.
- **Give permission to call out bad behaviours** and correct them immediately.
- **Encourage and reward behaviours and attitudes linked to D&I,** such as tolerance, empathy, open-mindedness, agility and flexibility, curiosity and looking beyond the label.
- **Develop HR policies that are family-friendly.**

On this note, my strong advice to any CEO is to create a workplace that is inclusive, but make sure that the culture throughout all levels is truly inclusive, and it is not just a direction from on top. Ensure all management levels share this ideal and practise that pursuit of inclusiveness. Where you are steering a large enterprise, realise the power you hold to create change and instil a culture that searches out diversity, makes all forms welcome, and builds a business structure that will thrive on inclusivity.

Antonia Belcher, Founding Partner, mhbc

(ii) Considering gender-specific issues

It is important to recognise first how important gender diversity is in the workplace. Helping encourage more women to enter the sector, and supporting those that do, is so important, especially for this sector where the percentages remain so low. This is not a case of supporting women over men, but providing a workplace environment where the differences each bring are recognised, nurtured and celebrated.

Of course women think differently to men, and having greater diversity at all levels in your organisations brings fresh perspectives, is likely to mirror your client's expectations, but will also deliver more valuable talent. . . .

Set the right culture. Engender a spirit of kindness. Train your managers. Seek out the talent in the organisation and promote it – but proactively look for and encourage more women to come forward.

The paybacks to organisations such as yours are immeasurable in providing a greater balance and enabling a better working environment for all.

**Pinky Lilani, founder and Chair of Women
of the Future Awards and Asia Women of
Achievement Awards**

- **Consider the gender agenda for your organisation.** What are your organisational policies regarding gender inequality and equality?
- **Consider what might work for your organisation to improve the gender balance** at all levels.
- **What can you do to actively encourage more applications from more women?**
- **What are the gender pay gap implications in terms of reporting requirements?**

- **Consider how can you tackle closing the gender pay gap through effective actions** by:

  o including multiple women on shortlists for recruitment and promotions;
  o using skill-based assessment tasks in recruitment;
  o using structured interviews for recruitment and promotions;
  o encouraging salary negotiation by advertising salary ranges;
  o introducing transparency to promotion and pay and reward processes;
  o appointing diversity managers and/or diversity task forces.

- **Promising actions** include:

  o improving workplace flexibility for men and women;
  o allowing people to work flexibly;
  o encouraging uptake of shared parental leave;
  o recruiting returners;
  o offering mentorship and sponsorship;
  o offering networking programmes;
  o setting internal targets.

- **Ensure gender equality is a given in your organisation**, from equal pay to considering the Sustainable Development Goal targets.
- **Create a female-friendly environment** that embodies aspects such as an inclusive culture, opening up relationships with senior leaders in the organisation who can act as mentors, providing access to challenging job assignments and enacting objective recruitment and promotion policies.
- **Help to promote the great female role models within your organisation.**
- **Encourage the development of, and support, female networks** to provide an environment for women to come together.
- **Consider how you can personally support women in the organisation and how you can encourage other leaders to become effective mentors.**
- **Encourage the women in your organisation to be the best that they can be and to aspire to senior positions** – not to the detriment of their male colleagues, but alongside them.

There are no superwomen out there: from interns to board level, we all get the same 24 hours every day, and we are just normal human beings working hard to keep everything together. Trust us, it is not always easy. The challenges are real, and so are the feelings of guilt and inadequacy – guilt for not always being there for our families, especially our children, guilt for leaving the office before some of our colleagues to see to our families,

(continued)

(continued)

and most of all, feeling guilty because we love what we do so much that sometimes we don't want to go home.

Some women have an even bigger challenge: being a single working parent with a limited support system. This working single parent is everything and all things to the family they are raising. A single parent is one of the most amazing employees an organisation can have, as they are extremely loyal, masters at multi-tasking, and hardworking because their family's livelihood depends on their income. Make sure you look after these employees, give them what they need to excel in their work environment, and remember that there is no "one size fits all". Flexibility is key: flexible working conditions will lead to a creative working environment, and time is more important to a single parent than just pure income. Trust them to deliver the high-quality job you employed them to do. They will repay your understanding and support tenfold. Have some empathy, understand that it is not always easy, and sometimes a smile and a thank you is all they need. The results will be overwhelming, and the output will increase substantially. Trust me.

Sanett Uys, Managing Director,
Serendipityremix

Development

(i) Developing your people

Consider your top employees right now: are they looking after themselves? Are they working at 100%? Would you even know if they weren't?

Take some time to ask. Show them some feeling and compassion in their own struggles. Invest some of their working hours into their wellbeing. Provide easy-entry solutions like meditation to begin the journey. It doesn't take much to transform those employees, their work and the overall culture into something that will pay dividends for everyone for years to come.

Gian Power, founder and CEO of TLC
Lions and Unwind London

- **Provide the right training and promotion policies** that offer equal access to career progression, enabling all staff to develop and succeed, including the following:

o Fair access to training and appraisals helps to ensure that organisations are developing and progressing their best talent.

o Pathways for requesting training must be transparent to all employees.

o It is important to take into account the accessibility of training locations, timing and delivery methods.

o Appraisals are essential, as is monitoring an individual's development through a personal development plan so that the impact on D&I can be better understood.

o Training managers about unconscious bias to reduce bias in appraisals and promotion choices is essential.

o Considering D&I goals when implementing fast-track leadership programmes should be adopted.

- **Actively consider setting targets, and monitor progress on key metrics.**
- **Put in place opportunities for promoting great role models across the organisation and beyond.** Enter people for awards, and promote a more D&I face in the public profile of your organisation.
- **Establish employee networks where people can informally come together** to support, coach or mentor each other.

(ii) Mentoring

- **Consider a mentoring, and reverse mentoring, programme** to help support and develop talent within the organisation:

  o Clear role models, with whom individuals from diverse backgrounds can identify, can inspire them to emulate the successes of their role models.

  o Where internal, diverse role models do not currently exist, organisations have invited external role models to speak about their experiences.

  o Role-modelling is not just about individuals' backgrounds – it is also about inclusive behaviour.

  o Leaders who exude inclusivity through their management style and consideration of individuals' needs can be powerful role models.

  o Lack of visible role models is cited as a reason by some job candidates for not engaging with the industry.

(iii) Staff engagement

- **Gain employee feedback,** as this is essential – the consistent use of staff surveys or forming employee groups to gain employee feedback are seen as effective mechanisms in common practice.
- **Having D&I champions to drive change at a local level** works well when coupled with senior-level engagement.
- **Building D&I awareness and training for all staff** provides them with an understanding of the skills and the language needed to help change the culture of their organisation and make it more inclusive.

(iv) Continuous improvement

- **Continuous improvement is not a "nice to have", but is essential** in ensuring the D&I agenda remains relevant to where the organisation is currently at and is reflective of what senor leaders and employees want.
- **Established action plans against D&I goals need to be regularly reviewed.**
- **Sharing and learning from leading practices**, both from within the sector and outside it, will help improve an organisation's D&I journey.
- **Professional organisations, such as RICS, are a great source of information and data.** They also provide a catalyst for learning and sharing information about D&I with others.
- **Work with other key influencers and stakeholders on the D&I agenda**, including international organisations, real estate industry bodies and branch organisations, D&I organisations and education providers, as all have a role to play and information to share.
- **Review and measurement of progress against D&I goals** allows the organisation to understand what it should start, stop or continue doing.

Here's what worked and what I believe we need to see more of to reflect the society the real estate sector serves:

- When planning recruitment, pause to think about the likely outcome of your approach . . . .
- Seek some help to understand unconscious bias that may exist in your organisation . . . .
- Cultivate a culture of being the "best version of yourself" – for everyone . . . .
- Avoid "one size fits all". Yes, we often need policies and governance to make things work effectively. But there are rules that by their nature can detract from the creation of a diverse workforce. Working hours, dress codes, the ability to work from home or remotely – all of these have a massive impact on your ability to attract, retain and engage.
- Create an open culture of questions and listening to understand . . . .

I've seen some glaring examples of non-diverse and uninclusive design recently. Prayer rooms at the far end of a corridor added as an afterthought. A lack of thought for nursing mothers in public buildings where they are core customers. Place-making where "shared space" schemes, where buses and pedestrians mix, scare parents of young children and the elderly. We know we have the power to shape our built environment. For me, there is no better reason than this to ensure we have the right people around the table in our organisations to help create a diverse and inclusive society.

Sue Willcock, Director, Chaseville
Consulting Ltd

8.6 Message to the CEO

Having vision and showing leadership as the CEO will, of course, not be new. But to truly make a difference when it comes to the D&I agenda, your role is critical. People within and outside the organisation will look to you to provide the aspiration and show the way.

The very fact that you are reading this book says that you already understand the importance of D&I and are already on the journey. This is such an important agenda in this current cycle of talent attraction and retention. Have the courage to move the dial for your organisation by increasing the emphasis and focus. This will leave a legacy for future generations.

Your role is to set the dial for change on D&I for your organisation, recognising what stage of the journey you are on and to:

Be the change that you want to create

"Dial set to DiversiTEA and INKlusion. So put this book down already, and go be the change you want to create."

The time is now. Diversity and inclusion in this sector is woefully behind many others, and yet the opportunities to attract, retain and grow talent have never been more important. With a well-publicised war for talent evident, people are invariably attracted to organisations that are seen to be positively creating an environment where everyone can be successful, can be themselves and can have a successful career.

Arun Batra, CEO/founder, UK National Equality Standard

The time for talk on D&I in the sector is over. It's now time for action. So here's a suggestion on a ten-point personal action plan.

1 Set a vision and strategy for the organisation on D&I, publish it, and talk about it with your employees – set aside resources (and budget) to support the agenda.
2 Be an authentic leader when it comes to D&I – do not pretend or say anything you do not believe in.
3 Set the language and tone by calling out openly bad behaviours, and neutralise language to be acceptable for everyone across the organisation – for example, stop calling a mixed group "guys".
4 Make a public pledge to refuse to speak on panels that are not diverse.
5 Ensure all your recruitment shortlists include more than one female, and are diverse wherever possible.
6 Operate a flexible working policy in your organisation, and practise it yourself, as well as rewarding output, not presenteeism. Discourage organisational flexism.
7 Bring in shared parental leave.
8 Set up and support networks in the organisation, or support your people in attending these outside the organisation.
9 Try to do more about attraction to the sector by working with schools and professions to communicate with the next generation.
10 Walk the talk – and speak about D&I in your organisation openly. Remember that sometimes it is the small things you do or say that will have the biggest impact.

And finally:

**Diversity and inclusion is the responsibility of everyone,
wherever you sit in the organisation.**

Make sure you are playing your part.

Note

1 Hunt, V., Prince, S., Dixon-Fyle, S. & Yee, L. (2018). *Delivering through Diversity*. New York: McKinsey & Company. Available from: www.mckinsey.com/~/media/mckinsey/business%20functions/organization/our%20insights/delivering%20through%20diversity/delivering-through-diversity_full-report.ashx [Accessed 30 September 2018].

Biographies

Amanda Clack MSc BSc PPRICS FRICS FICE FAPM FRSA CCMI FIC
CMC AffiliateICAEW

Amanda Clack is an Executive Director at CBRE, where she is Head of Strategic Advisory and Managing Director for Advisory across Europe, the Middle East and Africa (EMEA). She is a member of both the CBRE UK Board and the EMEA Occupier Board.

As the 135th President for the Royal Institution of Chartered Surveyors (RICS), Amanda Clack held office in 2016–17 for a period of nearly 18 months, and as such, became the longest-serving president in 123 years. During her presidency, her themes focused on Infrastructure, Cities and the War for Talent (skills, diversity and inclusion).

Amanda is listed in *Who's Who*, and is recognised as a senior property professional in real estate, infrastructure and construction, with a career spanning over 30 years. In addition, Amanda has been a partner in both EY LLP, where she was the Head of Infrastructure (Advisory), and at PwC LLP, where she was Head of Profession for programme management and also ran the south-east consultancy business. She is an experienced Global Client Service and Engagement Partner, so she understands the importance of people in business.

Her academic and professional qualifications include an MSc in Programme Management and BSc Quantity Surveying, plus she is a Fellow of the RICS, the Institution of Civil Engineers, the Association of Project Management, the Institute of Consulting and the Royal Society of Arts. Amanda is a Certified Management Consultant and Companion of the Institute of Management. She was given a Success in Business Award by Anglia Ruskin University in 2013, and has won numerous industry awards, including: the David Bucknall Award 2018 (for outstanding contribution to the quantity surveying profession globally), Woman of the Year 2015, and the Women in the City Woman of Achievement Award 2015 for Construction and Engineering, and was also pronounced as a Mover and Shaker of 2015 in the ICAEW's Global Finance 50, as well as having been named in City AM's Power 100 Women list 2016. Amanda has been nominated and included as part of the Pride in the Profession for the RICS 150th Anniversary.

In the UK, Amanda was also a member of the All Parliamentary Group for Excellence in the Built Environment (APGEBE) as part of a Commission of Inquiry into Skills.

Judith Gabler MSc BA (Hons) Dip RSA FCMI CMgr FCIL CL

Judith Gabler holds language qualifications in French and German from the University of Manchester, the University of Central Lancashire, the Royal Society of Arts and the Chamber of Commerce in Wiesbaden, Germany. She also holds an MSc in Real Estate and Property Management from the University of Salford. In addition, she is a Chartered Fellow of the Chartered Management Institute (CMI), as well as a Fellow and Chartered Linguist of the Chartered Institute of Linguists (CIOL).

Having worked in Germany for over 30 years and for RICS in Europe since mid-1995, Judith was previously the driving force behind developing RICS in Germany, and is currently Interim Managing Director, Europe. She has long-standing experience of managing pan-European teams, and at the heart of everything she does is caring and bringing out the best in everyone, taking into account all the different cultures, experiences and personalities.

Her volunteer commitments extend to various social projects, as well as mentoring in order to help younger people realise their career aspirations. She is also the Chair of the CIOL Council, an honour and responsibility she is committed to in order to promote language qualifications and standards worldwide, and ensure that languages remain an attractive and exciting global career choice.

In 2006 Judith was listed among the top 40 leaders in German real estate by the magazine *Immobilien Wirtschaft*.

In 2018 she was ranked the number one female in the category "Influencer" by *Immobilien Manager* magazine for her work promoting RICS, standards and values in the German real estate sector.

Judith has lived in Frankfurt since 1984 and holds both British and German citizenship. She is married, with two daughters.

Index

Page numbers in **bold** refer to tables.

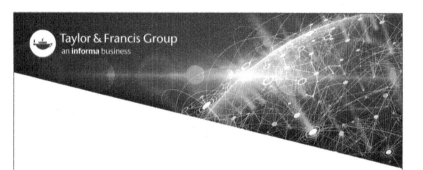